MILITARY SIMULATION

& SERIOUS GAMES

WHERE WE CAME FROM AND WHERE WE ARE GOING

ROGER D. SMITH

Author of *Game Technology for Medical Education*
and *Simulation Interoperability*

M

Modelbenders Press

The Library of Congress has cataloged the paperback edition as follows:

Smith, Roger
    Military Simulation & Serious Games: Where we came from and where we are going.
    Roger Smith.—1ˢᵗ ed.
      1. Technology—Computers   2. Education and Training
      3. Simulation and Games   I. Roger Smith   II. Title

ISBN-13: 978-0-9823040-6-8
ISBN-10: 0-9823040-6-4

TABLE OF CONTENTS

FOREWORD:
PANCAKE PEOPLE

Historically each individual has had the opportunity to become very deep in a very narrow part of existence. The farmer on an isolated piece of land in the Great Plains spent his entire life learning the fine details about the weather, soil, wildlife, growing patterns, and planting strategies that worked at a particular point of latitude and longitude. His education began as a boy between the ages of 5 and 8 years-old. If he chose to stay on the farm and carry on the family business, his experience and understanding of that point in time and space would continue until he was 60 or 70 years-old. This lifestyle allowed him to become an expert with fifty or sixty years of expertise in a very specific domain of industry, geography, and time. He had the opportunity to socialize with other farmers around him and to learn from their experience as well. Like a scientist who spends all of his time in personal experimentation and works only with a tight-knit group of kindred souls, this farmer was able to master one specific domain through hard work and endurance.

Similar pockets of focused expertise existed all across the country and the world. They were characteristic of the "up by your own bootstraps" determination that was required to survive and succeed in isolated conditions and with little access to outside information. But as mediums for communication spread and became more accessible, the farmer was able to learn about the ideas of others around the state, the country, and the world. He began to rely on the expertise and discoveries of people far from his point in space, people he had never seen, met, nor imagined. He did not have to understand how their ideas worked. He just had to know enough to implement them effectively. Knowledge came to him embedded in new kinds of seeds, new equipment, and new farming practices that he had not created himself. He began to explore beyond his own experience with nature, work, and society. He was once an intellectual pillar standing at a particular point in space and time. He was tall, certain of his knowledge, and a master of his place in the world. But the distribution of new products and the communication of new information allowed him to expand out and become more knowledgeable about the world. He was able to learn about and experiment with ideas beyond those tied to survival. He could take on hobbies and indulge his interest in machinery beyond farming. The farmer was becoming more broad and well-rounded. He was flattening out to cover more area and to rely on the expertise of others to perform his core function.

This picture of the lone farmer working to survive is similar to what has happened in all parts of society. The television and the Internet have played an important and unstoppable role in the flattening and broadening process. Since the 1950's, technology has broadened our exposure to information and thinned our specialization in one specific topic—especially concerning topics that are directly aligned with professional and economic survival. We have created pancake people who know a little bit about everything in the world, but not a great deal about any one thing. Some individuals and groups have remained dedicated to mastering their specialty. Professionals in law, medicine, engineering, philosophy, history, literature, and similar disciplines still prided themselves on their depth of thinking and their understanding of the great writings in their fields. However, the recent explosion of the World Wide Web has begun to link, summarize, critique, and repeat all knowledge. Even the islands of experts have found themselves seduced by easy access to vast stores of information. Even they have become information surfers. Though they may have criticized the masses of television channel surfers, they have succumbed to the same behavior. The seduction of surfing across all information in any field, the ability to quickly locate and scan anything that has been written, these opportunities are just too tempting to resist. But as we navigate these vast tracts of information we are just beginning to notice that the depth of our absorption, understanding, and incorporation of that information into our own ideas is diminishing. The intellectual classes of society are being broadened and flattened by the Internet in the same way that the television flattened the general public. Pancake intellectuals are joining the ranks of a pancake society.

From an intellectual perspective, the ability to think deeply about a subject may be attributed to the creation of the book. This abstract representation of knowledge could be patiently and painstakingly created by an author. It could be shared with millions of readers, who could spend their time working through it. Prior to its invention there was no means to collect, record, and disseminate large volumes of thought. The Internet is changing the era of the book into the era of the page. It is returning society to a time when all knowledge was limited to the size of a single piece of paper. It appears that we are not veering into a new type of information exchange, but are returning to a previous pattern.

Are we better off as pancake people?

Certainly exposing someone to a broad sampling of information in their early years is an advantage. It allows them to consider many options for directing their life. It opens doors that were previously closed or unimagined. But, is the shallow surfing of information the best way to conduct one's life indefinitely? It is an effective means of discovering a new field, but not of mastering it. Sergey Brin argues that people are better off if they have access to all of the world's information. That is certainly true. But it does not speak to the question of how a person should use that access. Being able to read the entire encyclopedia of human knowledge exposes the surface of a very large world. But it is not an effective method for becoming a master of any one part of that world. The expert must focus his attention on one small area for a significant amount of time. Nicholas Carr has asked whether it is possible to develop a mind which can both surf broadly and penetrate deeply into information. His own experience has been that broad surfing has weakened his own ability to penetrate deeply. It appears to be a choice that each person must make—the wide flat pancake of familiarity, or the deep-rooted pillar of expertise.

This book is an attempt to provide some depth of thought on a single subject. It seeks to contribute to the professional understanding of military simulation, serious games, and the technologies that support them. We want to contribute to the pillar of expertise in the face of the pancaking of people in every profession.

Part 1

MILITARY SIMULATION TECHNIQUES

Chapter 1

SIMULATION: THE ENGINE BEHIND THE VIRTUAL WORLD

*S*imulation is the process of designing a model of a real or imagined system and conducting experiments with that model. The purpose of simulation experiments is to understand the behavior of the system or evaluate strategies for the operation of the system. Assumptions are made about this system and mathematical algorithms and relationships are derived to describe these assumptions—this constitutes a "model" that can reveal how the system works. If the system is simple, the model may be represented and solved analytically. A single equation such as DISTANCE = (RATE * TIME) is an analytical solution representing the distance traveled by an object at constant rate for a given period of time.

However, problems of interest in the real world are usually much more complex than this. In fact, they may be so complex that a simple mathematical model can not be constructed to represent them. In this case, the behavior of the system must be estimated with a simulation.Exact representation is seldom possible in a model, constraining us to approximations to a degree of fidelity that is acceptable for the purposes of the study. Models have been constructed for almost every system imaginable, to include factories, communications and computer networks, integrated circuits, highway systems, flight dynamics, national economies, social interactions, and imaginary worlds. In each of these environments, a model of the system has proved to be more cost effective, less dangerous, faster, or otherwise more practical than experimenting with real system.

For example, a business may be interested in building a new factory to replace an old one, but is unsure whether the increased productivity will justify the investment. In this case, simulation would be used to evaluate a model of the new factory. The model would describe the floor space required, number of machines, number of employees, placement of equipment, production capacity of each machine, and the waiting time between machines. Simulation runs would then evaluate the system and provide an estimate of the production capacity and the costs of a new factory. This type of information is invaluable in making decisions without having to build an actual factory to arrive at an answer.

Simulations are usually referred to as either discrete event or continuous, based on the manner in which the state variables change. Discrete event refers to the fact that state variables change instantaneously at distinct points in time. In a continuous simulation, variables change continuously, usually through a function in which time is a variable. In practice, most simulations use both discrete and continuous state variables, but one of these is predominant and drives the classification of the entire simulation.

History. One of the pioneers of simulation concepts was John von Neumann. In the late 1940's he conceived of the idea of running multiple repetitions of a model, gathering statistical data, and deriving behaviors of the real system based on these models. This came to be known as the Monte Carlo method because of the use of randomly generated variates to represent behaviors that could not be modeled exactly, but could be characterized statistically. von Neumann used this method to study the random actions of neutrons and the effectiveness of aircraft bombing missions. These methods wee first used in industry to determine the maximum potential productivity of factories.

Concepts for Discrete Event Simulations (DES) were developed in the late 1950's. The first DES-specific language was developed at General Electric by K.D. Tocher and D.G. Owen. The General Simulation Program (GSP) was created to study manufacturing problems at General Electric and was shared with the rest of the world at the Second International Conference on Operations Research.

Purpose. Simulation allows the analysis of a system's capabilities, capacities, and behaviors without requiring the construction of or experimentation with the real system. Since it is extremely expensive to experiment with an entire factory to determine its best configuration, a simulation of the factory can be extremely valuable. There are also systems, like nuclear reactions and warfare, which are too dangerous to carry out for the sake of analysis, but which can be usefully analyzed through simulation.

Advantages and Disadvantages. When conducting a simulation or contriving a model, certain limitations must be acknowledged. Primary among these is the ability to create a model that accurately represents the system to be simulated. Real systems are extremely complex and a determination must be made about the details that will be captured in the model. Some details must be omitted and their effects lost or aggregated into other variables that are included in the model. In both cases, an inaccuracy has been introduced and the ramifications of this must be evaluated and accepted by the model developers. Another limitation is the availability of data for describing the behavior of the system. It is common for a model to require input data that is scarce or unavailable. This issue must be addressed prior to the design of the model to minimize its impact once the model is completed.

Both of the limitations listed above lead to a simulation that provides approximate results or that describes system behavior statistically. For this reason, simulation usually provides measurements of general trends, rather than exact data for specific situations or individuals. A simulation would be hard pressed to determine which piece of material will actually be ruined by a milling machine. But, it would be an excellent tool for determining the impacts of machine failure on factory productivity, using known statistical distributions for the failures of many machines.

COMMON APPLICATIONS OF SIMULATION

Simulation is used in nearly every engineering, scientific, and technological discipline. In the fifty years since its formal definition it has been adapted for a wide variety of applications. Today, the techniques are employed in the design of new systems, the analysis of existing systems, training for all types of activities, and as a form of interactive entertainment.

Design. Designers turn to simulation to allow them to characterize or visualize a system that does not yet exist and for which they wish to achieve the optimum solution. Manufacturing models may describe the capacities of individual machines, the time to prepare material for operation, time to transfer materials from one machine to another, the effects of human operators, and the capacities of waiting queues and storage bins. Simulations of new pieces of equipment may evaluate their performance, stress points, transportability, human interfaces, and potential hazards to the environment. Business process models may evaluate the flow of paperwork through a company to determine where redundancies or unnecessary operations are located, allowing them to redesign operations such that the same work can be performed with a fraction of the labor and time that has evolved into the process. Major airlines use simulations to study complex routing patterns for large numbers of aircraft traveling around the world. The intention is to identify routes that serve the most passengers and use the fewest assets most efficiently. Factors such as aircraft capacity, ground time, flight time, scheduled maintenance, crew availability, weather effects, and unscheduled downtimes are all considered in such models.

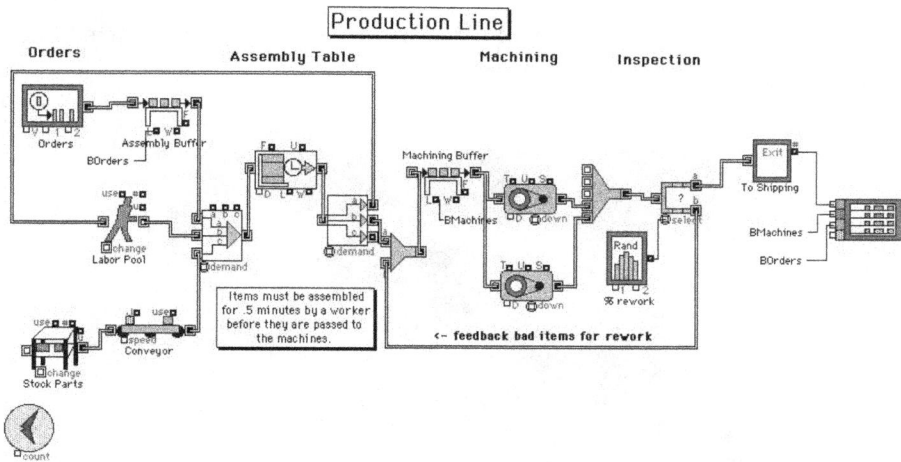

Simulation of a Manufacturing Production Line
Courtesy of Imagine That Inc.

Analysis. Analysis refers to the process of determining the behavior or capability of a system that is currently in operation. Unlike design, analysis may be supported by the collection of data from the actual system to establish model behaviors. The model can then be modified to determine the optimum configuration or implementation of the real system. A computer network can be described by the volume of traffic carried, the capacity of the lines and switches, performance of a router, and the path taken from sender to receiver. Based on measured message patterns, the network can be configured to deliver the most information using the shortest or most reliable paths available. In the health care industry, the models are used to schedule doctors, staff, equipment, and patients in an effort to improve service times and reduce costs. Social trends can be simulated to determine what services or goods will be needed at a given time by a specific sector of society. The impacts of aging, health, family composition, and a host of other factors can be predicted from an appropriate social model.

Traffic Flow Simulation
Courtesy of Intergraph Computers Inc.

Training. Training simulations recreate situations that people face on the job and stimulate the trainee to react to the situation until the correct responses are learned. These devices produce experienced personnel without the expense of making mistakes on the job. Perhaps the best known of these are flight simulators, which model dangerous environments where life threatening situations can be mitigated through learning in a non-lethal environment. Military simulators replicate the performance characteristics of the aircraft, instruments in the cockpit, effects of weapons, support from other combat systems, communications with other pilots, and terrain over which the events occur. Similar systems are used to train the captains of large ocean-going ships to dock without destroying both a real ship and a real dock. Entire mock-ups are made of nuclear power control centers to teach operators how to respond to emergency situations and to identify potential hazards before a crisis occurs. Modern medical equipment is so expensive and scarce that simulations have been constructed to allow interns and nurses to practice, develop, and certify their skills without having to schedule training time on the real equipment, competing for its use by real patients.

Flight Simulator
Courtesy of SEOS International Ltd.

Entertainment. The entertainment industry makes wide use of simulation to create games that are enjoyable and exciting to play. These contain many, but usually not all of the components of simulation described in this article. Arcade games, computer games, board wargames, and role playing games all require the creation of a consistent model of an imaginary world and devices for interacting with that world. These simulations often appear very similar to training simulations, but differ in that their purpose is entertainment rather than practice for real-world events. This fact allows game developers the freedom to modify the laws of physics and other behaviors, rather than accurately capturing their real world equivalents. Advances in these simulations, together with the prevalence of the Internet, are allowing the creation of multi-player on-line games that pit players against multiple opponents around the world. Though the purpose of these simulations is entertainment, the technical challenges faced by their developers are just as daunting as those in the other categories.

Tank Simulator Computer Game
Courtesy of MaK Technologies, Interactive Magic, and Zombie Studios

All of the areas listed here allow systems to be understood without incurring the expenses or dangers of working with the actual system. As the benefits of simulation become more widely understood, and the complexity of modern problems increases, the user base for simulation will grow rapidly.

THE SIMULATION DEVELOPMENT PROCESS

The creation and operation of a simulation was once a black art in which only experienced practitioners could claim competence and understanding. However, over the last several decades a definite process has evolved for developing, validating, operating, and analyzing the results of simulations. In this section we will describe the process illustrated in Figure 1.1.

Define Problem Space. The first step in developing a simulation is to explicitly define the problem that must be addressed by the model. The objectives and requirements of the project must be stated along with the required accuracy of the results. Boundaries must be defined between the problem of interest and the surrounding environment. Interfaces must be defined for crossing these boundaries to achieve interoperability with external systems. A model can not be built based on vague definitions of hoped for results.

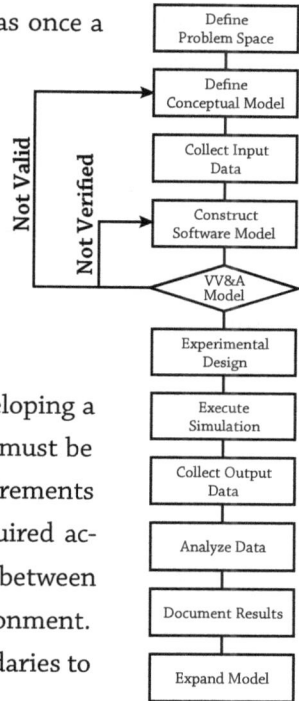

Figure 1.1. Simulation Development Process

Define Conceptual Model. Once the problem has been defined, one or more appropriate conceptual models can be defined. These include the algorithms to be used to describe the system, input required, and outputs generated. Assumptions made about the system are documented in this phase, along with the potential effects of these assumptions on the results or accuracy of the simulation. Limitations based on the model, data, and assumptions, are clearly defined so that appropriate uses of the simulation can be determined.

The conceptual model includes a description of the amount of time, number of personnel, and equipment assets that will be required to produce and operate the model. All potential models are compared and trade-offs made until a single solution is defined that meets the objectives and requirements of the problem and for which algorithms can be constructed and input data acquired.

Collect Input Data. Once the solution space has been determined, the data required to operate and define the model must be collected. This includes information that will serve as input parameters, aid in the development of algorithms, and be used to evaluate the performance of the simulation runs. This data includes known behaviors of working systems and information on the statistical distributions of the random variates to be used. Collecting accurate input data is one of the most difficult phases in the simulation process, and the most prone to error and misapplication.

Construct Software Model. The simulation model is constructed based on the solution defined and data collected. Mathematical and logical descriptions of the real system are encoded in a form that can be executed by a computer. The creation of a computer simulation, as with any other software product, should be governed by the principles of software engineering.

Verify, Validate, and Accredit the Model. Verification, validation, and accreditation (VV&A), is an essential phase in ensuring that the model algorithms, input data, and design assumptions are correct and solve the problem identified at the beginning of the process. Since a simulation model and its data are the encoding of concepts that are difficult to completely define, it is easy to create a model that is either inaccurate or which solves a problem other than the one specified. The VV&A process is designed to identify these problems before the model is put into operation.

For the purposes of VV&A the simulation development process is divided into the problem space, conceptual model, and software model with definite transitions and quality evaluations between these stages as shown in Figure 1.2. Validation is the process of determining that the conceptual model reflects the aspects of the problem space that need to be addressed and does so such that the requirements of the study can be met. Validation is also used to determine whether the operations of the final software model are consistent with the real world, usually through experimentation and comparison with a know data set. Verification is the process of determining that the software model accurately reflects the conceptual model. Accreditation is the official acceptance of the software model for a specified purpose. A software model accredited for one purpose may not be acceptable for another, though it is no less valid based on its original design.

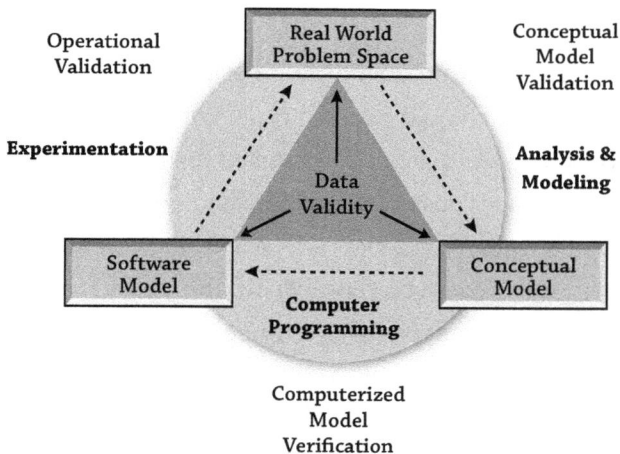

Figure 1.2. VV&A Process

Design Experiments. This phase identifies the most productive and accurate methods for running the simulation to generate the desired answers. Statistical techniques can be used to design experiments that yield the most accurate and uncompromised data with the fewest number of simulation runs. When simulation runs are expensive and difficult to schedule, experimental design can ensure answers at the lowest cost and on the shortest schedules.

Execute Simulation. This is the actual execution of the designed, constructed, and validated model according to the experimental design. The simulation runs generate the output data required to answer the problem initially proposed. In the case of Monte Carlo models, many hundreds or thousands of replications may be required to arrive at statistically reliable results.

Collect Output Data. Concurrent with the execution of the model, output data is collected, organized, and stored. This is sometimes viewed as an integral part of the model, but should be distinctly separated since it is possible to change the data collected without changing the model algorithms or design.

Analyze Data. Data collected during the execution of a simulation can be voluminous and distributed through time. Detailed analyses must be performed to extract long-term trends and to quantify answers to the driving questions that motivated the construction of the simulation. Analysis may produce information in tabular, graphic, map, animation, and textual summary forms. Modern user interfaces have greatly enhanced this phase by displaying data in forms that can be easily understood by diverse audiences.

Document Results. The results of the simulation study or training session must be documented and disseminated to interested parties. These parties identify the degree to which the simulation has answered specific questions and areas for future improvements.

Expand Model. Simulation models are expensive and difficult to build. As a result, once a model is built, it will be modified for use on many related projects. New requirements will be levied, new users will adopt it, and the entire development process will be conducted many times over.

FUNDAMENTAL TECHNIQUES FOR MODEL BUILDING

Though simulations vary widely in their design and implementation, most share a few common features to achieve their goals.

Event Management. A simulation is made up of states, events, and entities. States are groups of variables that describe the system at a specific time. Events are activities that change the state of the system. Entities are the objects represented in the simulation, the things described by the state variables and to which events occur. Events are the key items that make transformations in the model and drive it through its operations. These may include the arrival of a piece of material at a milling machine, the departure of an aircraft from an airport, the delivery of a message in a network, or an engagement between a missile and a fighter aircraft. These are typically managed through the use of multiple lists or queues in the model. The queues identify which events are ready to be processed, which are waiting until a specified time, and which must be triggered by specific conditions.

Queues manage events by ordering and releasing them according to specified criteria. The most common type of queues are the First In First Out (FIFO), Last In First Out (LIFO), Ordered, and Random. Each of these releases events into the simulation using a different method and each is uniquely useful for representing specific situations. The FIFO queue may contain a plan in the form of events to be executed. The LIFO queue may handle object reactions that interrupt and supersede planned events. An Ordered queue is widely used in training simulations, in which the time an event occurs drives its insertion into the simulation. A random queue assigns no order to the events in the queue, processing any one without regard for its priority or arrival order.

Time Management. In a simulation, the advancement of time is performed using a variable that can be controlled as any other and need not be tied to the advancement of real time or the internal computer clock. Typically, simulations move forward through the use of event based time (event stepped) or incremental time advancement (time stepped). An event-stepped simulation recognizes that in the model of the system, the only changes occur at the points at which events occur. Therefore, the model jumps from one scheduled event to the next, omitting the representation of intermediate times and speeding up the execution of the simulation by eliminating operations that do not effect simulation state. Time stepped simulations, on the other hand, are used when there are a large number of interactions between entities based on shared events.Training simulations use this method because of the need to present a consistent flow of time and events to a person that is interacting with the simulation.

A great deal of work has been done in the area of managing time stepped simulations. These simulations are discrete, stepping from one time to another using a specified increment. This increment may be fixed such that each step is one minute, one second, or one microsecond long, or it may be variable such that the size of the step is determined by the activity being simulated. It may be necessary, for example, to represent time in sub-seconds during a combat engagement and in days during the political negotiations prior to the beginning of hostilities.

As simulations have grown to operate across networks of computers or on parallel computers, the models have been separated into pieces that represent a portion of the problem, and exist as multiple software programs on multiple machines. This has resulted in the need to maintain consistency among programs, which can not be done effectively with the simple queuing lists used within a single program. Parallel and distributed time management was initially achieved through the use of a shared clock to which all programs would refer. However, research has lead to the use of algorithms that can ensure time synchronization without the use of a central shared clock.

Parallel and distributed time management can be accomplished through conservative or optimistic synchronization. Both methods use a mechanism to understand the time of each of the processes. Conservative synchronization then chooses to maintain consistency among all processes as the simulation executes. Optimistic synchronization, on the other hand, allows each process to move ahead as fast as computationally possible, putting each process at a different point in time. When an event is received from another process that affects past events in the local process, the simulation reverses its operations and "rolls back" in time to include the new event interaction. The conservative method assumes that interactions between processes are common enough that constant synchronization is the most efficient method of proceeding into the future. The optimistic method assumes that interactions are scarce and the problem can be solved most efficiently by working as fast as possible. These roll back only when interactions are received in the past.

Random Number Generation. Many models require the use of random numbers to introduce the variability caused by statistical rather than deterministic representations of events. Random number generators are used in the computers to replicate the creation of a series of numbers that are random and independent of each other. These algorithms are actually deterministic and merely provide the impression of randomness. Algorithms are typically required to be repeatable, fast, use little storage space, and usually generate Uniformly distributed numbers in the range of (0,1). To create variates from other distributions the Uniform random number becomes the input to a second algorithm that generates Normal, Exponential, Poisson, Gamma, Weibull, Lognormal, Beta, Binomial, or other distributed numbers.

Physical Modeling. Traditionally, models have represented the capabilities of machinery and systems based on their physical characteristics and the basic laws of physics. The focus has been on understanding and representing the physical environment—distance, rate, weight, density, etc. In manufacturing systems, models represent entities entering a system in which events are generated by statistical distributions buffered by waiting queues. In analytical physics simulation, the models represent the specific behaviors of particles or chemicals under specified conditions. In training simulations, the models reproduce the physical world allowing people to interact with terrain, buildings, and other entities. Each of these takes a unique view of the essential variables and algorithms needed to represent the physical behavior of the system. The impact of human decision-making and process variance is handled through the use of statistical distributions that represent variation with the aid of random number inputs to these distributions.

Behavioral Modeling. As the role of simulation has grown, the need to more accurately represent human and group behaviors has increased. To accommodate these needs, simulation developers have turned to the artificial intelligence community for assistance. Models now contain finite state machines, expert systems, neural networks, and case-based reasoning to represent human behavior in finer detail. This acknowledges that specific physical conditions trigger human behaviors that must be explicitly modeled to achieve the appropriate results. Behavioral modeling has been particularly useful in training applications where computer controlled adversaries with challenging and realistic behaviors are required.

Model Management. A computer simulation is a system of software and hardware that must be developed and managed in accordance with the same principles of systems and software engineering that govern other applications. Issues that are not germane to the **science** of simulation are very important to the **business** of simulation. An attractive and friendly user interface is important. Systems that have provided only textual input/output are giving way to those with graphical user interfaces and multi-dimensional representations of the simulated world. Configuration management of the models provides stability to a simulation program, ensuring that it can control its own evolution. Documentation provides permanence of expertise by recording model assumptions, algorithms, data collection, and validation results. This establishes a foundation that can extend the useful life of a model beyond the tenure of its original developers. Finally, domain architectures are being developed which attempt to capture the essential components and interactions of entire families of simulations. The intent is to create a structure that eliminates redundant development and promotes the reuse of modules or components that provide common functionality and interfaces across an entire family of simulations.

ESSENTIAL COMPUTER TECHNOLOGIES

Simulations, like all other applications, leverage technologies from other areas of science. The algorithms and information required to create a very complex model usually exceed the power of the available computer hardware and software necessary to run it. However, simulation programs are growing larger and more useful as a direct result of advancements in computer science. A few of the most useful technologies are described here.

Networks. The ability to distribute a simulation across a network of computers leads to more detailed, scaleable, complex, and accessible models. Distributed message passing and event synchronization allow a single problem to be addressed with a large number of traditional computers on a network. The proliferation of standardized networks between computerized machinery, communications systems, decision aids, and other tools has created an environment in which simulations can drive "real world" computers directly and extract data from them in real time. This has blurred the boundary between real and simulated worlds.

Parallel Computing. Parallel computing provides many of the advantages of networked computers, but adds the characteristic of close coupling. Some problems can be divided into many thousands of separate processes, but the interactions between them are so frequent that a general-purpose network for delivering messages introduces delays that greatly extend the execution time of the simulation. In these cases, parallel computers can provide the close coupling between processors and memory that allow the simulation to execute much more quickly.

Artificial Intelligence. The representation of human and group behavior has become essential in some parts of the simulation community. Techniques developed under the umbrella of artificial intelligence and cognitive modeling can solve some of these problems. Simulations are including more finite state machines, expert systems, neural networks, case based reasoning, and genetic algorithms in an attempt to represent human behavior with more fidelity and realism.

Computer Graphics. Simulation data lends itself very well to graphic displays. Factories and battlefields can be represented in full 3D animation using virtual reality techniques and hardware devices. Graphical user interfaces provide easy model construction, operation, data analysis, and data presentation. These tools place a new and more attractive face on simulations that previously relied on the mind's eye for visualization. This often leads to greater acceptance of the models and their results by the engineering and business communities.

Databases. Simulations can generate a large amount of data to be analyzed and often require large volumes input data to drive the models. The availability of relational and object oriented databases has made the task of organizing and accessing this information much more efficient and accessible. Previously, model developers were required to build their own storage constructs and query languages, a distraction from the real focus of the simulation study.

Systems Architecture. Simulations can be grouped into families, or domains, where the same software architectures can be used to model entire classes of problems. This recognition in transaction-based simulation has lead to the creation of a host of simulation products that encapsulate functionality used to model everything from factory operations to aircraft routing schedules.

World Wide Web. The expansion of the Internet and the World Wide Web has led to experiments with simulations that are either distributed through the Internet or accessible from it. These simulations make use of standard protocols and allow the distribution of a simulation across multiple computers that are not directly controlled on a dedicated network. Simulation users do not necessarily need to own the computers that run the simulation. Instead, the user may access a simulation-specific machine connected to the Web, provide input values, control model execution, and receive the results without ever having their own copy of the simulation software or the computers necessary to run it.

SIMULATION LANGUAGES AND TOOLS

A number of simulation languages and tools have been developed specifically to assist developers in constructing models of their systems. These languages are intended to serve a specific problem domain, rather than support general purpose programming as do FORTRAN, C, Pascal, and Ada. However, general purpose languages are still widely used to construct simulations in domains for which simulation specific languages or packages do not yet exist or where the problem is so unique that simulation tools can not be created economically.

Some of the more popular languages and tools are described below and a sample comparison of three provided in the table.

Discrete Event Simulation. Discrete event simulation includes a wide array of both problems and commercial tools for solving them. The descriptions below are separated into those that use an actual programming language and those that are simulation applications or toolkits.

Languages

Simula, the first simulation-specific programming language, was developed by O.J. Dahl and K. Nygaard of the Norwegian Computer Center. In 1961 they created Simula I by making extensions to ALGOL 60. The language was a general purpose programming language that included specific extensions to support simulation applications. Simula 67 was the first object oriented programming language, providing support for objects, classes, inheritance, encapsulation, multi-threading, and garbage collection. It was the motivation for the later development of the C++ language.

GPSS/H from Wolverine Software is a block programming language improved from the original GPSS developed at IBM in 1969. This language provides an interactive debugging environment, a floating-point clock, and built in mathematic, trigonometric, and statistical functions. It automatically collects basic simulation output data and supports the extension of this collection by the programmer.

SIMSCRIPT II.5 from CACI Products is an event-oriented and process-oriented language that evolved from the original SIMSCRIPT developed at the Rand Corporation in 1962. The language is actually a complete general programming language that can be used to build discrete-event, continuous, and combination simulations. It is supplemented by SIMGRAPHICS that allows the user to develop input forms, output displays, and interactive controls for the simulation.

SIMAN/Cinema from Systems Modeling is a combined simulation language and animation system. SIMAN models are constructed graphically using the Cinema package and automatically converted into code. The language includes built-in functions for manufacturing and material handling systems, an interactive debugger, and analyzers for input and output data.

SLAM II from Pritsker Associates is predominantly used for process-oriented simulation, with extensions that support event-oriented simulation and combinations of the two. The language represents models in a network-like structure that includes nodes and branches. Support packages allow the developer to draw a network, which is then converted into the simulation code.

MODSIM from CACI is an object-oriented programming language with graphic extensions to support data input, execution monitoring and control, and output analysis. The language includes built-in routines for statistical distributions and simulation management operations. Interaction with the language is through a development environment that includes a compiler, object manager, and debugger.

Language Comparison

To illustrate the syntax of some of the more common simulation languages, Table 1.1 provides the code for the same problem in GPSS/H, SIMAN, and SLAM II. This code represents a simple single-server queue with exponential inter-arrival and service times, such as a barbershop with one barber and a waiting queue. Though these languages are designed to efficiently serve the needs of model builders, their syntax is often equal or more complex than general purpose programming languages. This fact has motivated the creation of the simpler graphic packages and toolkits described in the next section.

Table 1.1. Simulation Language Comparison

GPSS/H	SIMAN	SLAM II
SIMULATE	BEGIN;	GEN, 1,,,,,,72;
GENERATE RVEXPO(1, 1, 0)	CREATE,,EX(1,1) :EX(1,1) :	LIM,1,1,100;
QUEUE SERVERQ	MARK(1);	NETWORK;
SEIZE SERVER	QUEUE, 1;	RESOURCE/SERVER(1),1;
LVEQ DEPART SERVERQ	SEIZE :SERVER;	CREATE,EXPON(1.0,1),1,1;
TEST L N$LVEQ, 1000, STOP	TALLY :1, INT(1);	AWAIT(1),SERVER;
ADVANCE RVEXPO(2, 0.5)	COUNT :1,1;	COLCT,INT(1),DELAY IN
STOP RELEASE SERVER	DELAY :EX(2,2);	QUEUE,,2;
TERMINATE 1	RELEASE :SERVER :DISPOSE;	ACTIVITY,EXPON(0.5,2),,DONE;
	END;	ACTIVITY,,,CNTTR;
START 1000		DONE FREE,SERVER;
END		TERM;
		CNTR TERM,1000;
		END;
		;
		INIT;
		FIN;

Tools

Extend from Imagine That is a visual, interactive simulation package for discrete event and continuous modeling that allows users to build models and user interfaces graphically. Model execution is carried out interactively on the graphic model representation. The package can accept data input through the interfaces or from separate files. Extend provides built-in mathematic and statistical functions and can be customized through the addition of C and FORTRAN routines. It also supports a client/server relationship with spreadsheet programs.

Workbench from SES is a visual simulation environment that allows models of complex systems to be built and executed graphically for performance analysis and functional verification. A model is specified graphically, as a hierarchy of directed graphs; declaratively, by filling in forms attached to each node in a graph; and procedurally, by specifying procedural methods attached to the nodes where desired in an internal language that is a superset of C.

TAYLOR II from F&H Simulations is a graphic model building package based on four fundamental entities—elements, jobs, routings, and products. These are manufacturing oriented where elements can represent machines, buffers, conveyors, transport, paths, warehouses, and reservoirs. The three basic operations supported are processing, transport, and storage. During simulation execution, graphic interfaces provide 2D and 3D views of factory activities.

COMNET III from CACI Products is designed to simulate communications networks. It provides a graphic interface for model building, execution, and data analysis. It specifically provides statistical distributions and control data for communications and computer networks as used by telephone companies, cable television broadcasters, and computer networks.

BONeS Designer from the Alta Group models the protocol and messaging layers of computer architectures and communications systems. The tool provides graphical user interfaces for defining data structures, analyzing results, generating finite state machines, and directing interactive simulation runs.

CSIM18 from Mesquite Software is a library of classes, functions, procedures, and header files that describe the activities and statistical distributions of communications, transportation, microprocessors, and manufacturing systems. Library components can be combined with developed software in C and C++ to create a simulation model that has fast execution.

SimPack from the University of Florida is a toolkit written in C and C++ to support the development of simulation programs by the user. It contains routines to support basic simulation operations and management of declarative, functional constraint, and combination models. The software can be combined with software written by the user.

CPSim from BoyanTech provides an execution kernel that manages synchronization, scheduling, deadlock prevention, and message passing, as well as a library of C functions that can be used to build an application. CPSim represents the system being modeled as a directed graph of communicating objects that are categorized as sources, nodes, and sinks. The tool supports portable models across single and multi-processor computers.

Continuous Simulation. The Advanced Continuous Simulation Language (ACSL) from MGA Software was developed specifically for modeling time-dependent, nonlinear differential equations and transfer functions. The language allows the user to create code from block diagrams, mathematical equations, and FORTRAN statements.

The Continuous System Modeling Program (CSMP) is constructed from three general types of statements—structural, which define the model, data, which assign numerical values to parameters, and control, which manage the execution of the model.

Interactive Simulation. In the interactive training arena a number of simulation products have emerged, particularly with military domain applications.

VRLink from MAK Technologies supports network protocols and simulation management for distributed military simulations. This package provides routines that format messages according to defined standards and manage the delivery and receipt of these messages across a number of computer platforms.

ITEMS from CAE Electronics provides a graphic environment for constructing simulated virtual worlds and the entities that populate them. The tool allows the creation of vehicles, aircraft, and humans and the specification of the physical characteristics and behavioral patterns. Terrain and weather data can be imported from standard formats or generated internally to create an operational environment. It provides a graphical user interface for executing and managing simulation runs.

MultiGenII from Multigen Paradigm is a three dimensional modeling tool for generating the visual representations of simulated objects, terrain, and cultural features for a complete synthetic environment to support training simulations. The tool simplifies the creation of the visual objects, allowing simulation developers to focus on more specific physical and behavioral models within the simulation.

CONCLUSION

Simulation Expansion and Transformation. Like all computer applications, modeling and simulation is expanding as a result of improvements in computer hardware and software technologies. There was a time when simulation was performed entirely by dedicated personnel using expensive, dedicated computer systems. We have reached a point where significant simulations can be performed on personal computers by experts in a specific field, without the need for a staff of simulation specialists.

Research in simulation itself is leading to an array of new technologies and methods for constructing and using models. Innovations include formalisms for defining models, interoperability of a diverse set of interactive simulations, metamodeling, human behavior modeling, and concurrent simulation.

Future. The manufacturing, research, planning, and training communities have discovered that answers to their questions and insights into their problems can be obtained economically and quickly from simulation models. As the world evolves into an information society, more and more business, recreation, and government activities will be defined in the form of digital data which can be organized, analyzed, and predicted using simulation. This power will drive the wide adoption of simulation by all forms of business and government.

Trademarked products referenced in the "Simulation Languages and Tools" section:

Trademark	Trademark Holder
GPSS/H	Wolverine Software
SIMSCRIPT II.5	CACI
SIMGRAPHICS	CACI
SIMAN/Cinema	Systems Modeling Corp.
SLAM II	Pritsker Associates
MODSIM	CACI
TAYLOR II	F&H Simulations
COMNET III	CACI
BONeS Designer	The Alta Group
CSIM18	Mesquite Software
CPSim	BoyanTech Inc.
ACSL	MGA Software
VRLink	MAK Technologies
ITEMS	CAE Electronics
MultiGenII	MultiGen Corp.

Chapter 2

MILITARY MODELING

T he military has always been a very heavy user and innovative developer of modeling techniques and technologies. The nature of military missions requires that they rehearse missions in order to better understand their complex interactions and to estimate outcomes. This need has led them to apply modeling and simulation to a number of different activities over the last 300 years. In this chapter we will explore the major applications of military modeling and will discuss the most come forms of dynamic modeling.

APPLICATIONS

The United States military has made its own unique definitions of the terms "modeling" and "simulation." For their purposes, modeling is often defined as, *"a descriptive, functional, or physical representation of a system"* (NSC, 2000). These representations may take the form of a mathematic equation, a logical algorithm, a three-dimensional digital image, or a partial physical mock-up of the system. Models are applied so widely that the variety of systems of interest is almost without bounds. In these systems military weapons systems are usually very prominently represented, to include land, air, and sea vehicles; communications and radar equipment; hand-held weapons; and individual soldiers. But models also represent the decision-making process and automated information processing that occurs inside the human brain and within battlefield computers. They extend to representations of the environment that is made up of terrain, vegetation, cultural features, the atmosphere, ocean, and RF environment. Different combinations of these are needed in order to accurately represent potential military situations.

One military definition of simulation is, *"a system or model that represents activities and interactions over time. A simulation may be fully automated, or it may be interactive or interruptible"* (NSC, 2000). This definition attempts to encompass human-in-the-loop simulators for training, as well as systems that serve as analytical tools for computing outcomes without the aid of a human participant.

The official categorization of the use of models and simulation within the military is to divide them into three large application groups.

The first is for use in "requirements and acquisition". In these applications, models are used to provide insight into the cost and performance of military equipment, processes, or missions that are planned for the future. These use scientific inquiry to discover or revise facts and theories of phenomena, followed by transformation of these discoveries into physical representations.

The second category is in exploring "advanced concepts". These models present military systems and situations in a form that allows the military to conduct concept exploration and trade studies into alternatives. These trade studies often explore multiple variations on a new weapon or tactic and attempt to measure the effectiveness of each of them. The result is a general appreciation for the different options available and some rough measure for ranking them. The models may be used to understand physical weapons or equipment, but they may also explore different processes for organizing and executing a mission. These require an understanding of processes and the interactions that occur between different steps in their processes. The models assist the military in creating its doctrine of operations, constructing its internal organization, and selecting materials for acquisition.

The third category is in "training and development". Models that are embedded in a simulation system are used to stimulate individuals and groups of personnel with specific military scenarios. The goal is to determine the degree to which they have learned to execute the doctrines they have been taught. It also gives them the opportunity to experiment with new ideas and to determine how useful these might be in a real warfighting situation. All of this can be done in a controlled environment that is free of life threatening situations that are part of real combat operations.

Finally, it should be noted that military modeling and simulation has always been the basis for a large segment of entertainment products. Many of the modeling concepts behind paper board-wargaming in the 1950's were developed simultaneously by the RAND Corporation for serious military training and by Charles Roberts at the Avalon Hill game company for popular entertainment (Perla, 1990). This trend has continued for over fifty years and can be seen today in comparing realistic three-dimensional military training systems and the very active computer gaming industry. Systems like America's Army provide an environment for experimentation and training in the military, a device to enhance Army recruitment and education about the military lifestyle, and a game for use by anyone looking for a little excitement in their free time.

REPRESENTATION

Over the past several decades, a number of different types of models have been developed for representing a military system or mission. These have gradually converged into commonly recognized categories of representation. These categories have significantly improved the ability of military modelers to communicate with each other and to exchange models without misunderstanding the differences between the products being created.

Engineering

Engineering models focus on the details of what a system does. These capture the physical properties of materials, liquids, aerodynamics, servomechanisms, and computer control of specific systems. They also include interactions between two physical objects or between an object and its environment. An engineering model attempts to understand the physical capabilities of the system at a level that is accurate enough to be used to design the system. Historically, physical prototypes were used to conduct these experiments. However, advanced computer technologies and modeling techniques have allowed us to create digital models of systems that are nearly as predictive as are live physical tests. These models offer many advantages over their physical counterparts. They are almost infinitely malleable so that experiments can be conducted on many thousands of variations rather than just a few physical prototypes. They are nearly infinitely instrumentable. It is possible to collect data from all points in space and time around the event of interest. When using physical prototypes we are often limited by our ability to place sensor, communication, and recording equipment at the precise place and time of interest.

Virtual

A "virtual model" often refers to a three-dimensional representation of a system that is operating in a digital three-dimensional environment. The focus is usually on the visual appearance of the object and the environment, more than on the properties of physics that were the focus of engineering models. Because of its visual focus, the objects most often represented are military vehicles and humans that would appear on a battlefield. This category is closely aligned with the more popularly recognized term, "virtual reality".

A virtual model and environment are usually constructed to simulate individual soldiers who are immersed in a system that generates visual, aural, and tactile stimuli. The goal is usually to train, test, or measure the ability of the human to respond in a desirable manner to the stimuli. Flight simulators are the most popularly recognized form of these models and systems.

Constructive

A "constructive model" represents an accumulation or aggregation of a number of objects, behaviors, and properties. In order to deal with the incredibly large and complex missions of the military, a very structured organizational hierarchy has evolved. To represent the information that is available at the different levels in this hierarchy and to represent the functions of the hierarchy itself, constructive models have been created.

A constructive model may represent a flight of four aircraft as a single item in the simulation. It may also group several hundred vehicles, humans, and equipment into a single object model. This model must then represent the aggregated behaviors of its many different constituent parts. There are a number of motivations for this type of modeling. First, it allows the simulation system developers to capture the operations of a much broader battlefield in a form that can be run on a reasonable computer suite. Second, in many cases the behavior of groups of objects are not understood at the engineering or virtual level, but can be represented as a higher-level aggregate. Third, this type of model mimics the organization, representation, and information that are used in the real military organizational hierarchy.

Very basic constructive models of military operations can be seen in many board and computer games, such as Chess, Stratego, and Risk. Constructive simulation systems differ from virtual systems in that the human operator or player is often positioned outside of the battle. Engagements are not usually targeted at the human player, so they are in a position to think more strategically about the situation and are not required to react to individual events that appear to threaten them personally, as would occur in a virtual system.

Live

Though a "live model" appears to be an inappropriate description, the term has been adopted to refer to activities in which live humans, vehicles, and equipment engage in mock combat. The combat events do not involve real munitions and attempt to avoid situations that could have lethal outcomes. Using computer, communication, navigation, and laser technologies, training areas have been constructed in which combatants can use their real weapons in a form that is as physically realistic as possible. Laser beams often replace bullets and computer messages indicate where bombs are dropped.

Live modeling allows humans to train in the real environment, to experience the physical hardships of traversing rough terrain, operating in the desert sun, and experiencing the effects of dirt and water on the equipment. The humans and vehicles become living models in a living simulation. In many cases, these live participants are also supplemented with virtual and constructive models to enrich the entire training experience.

Environment

The model of the environment has historically been a static representation of terrain, vegetation, roads, rivers, wind, clouds, rain, ocean waves, salinity, ocean bottom, and any number of other features. This environment has provided a medium within which the above models operated. The environment impeded the movement of objects, obstructed sensor visibility, and changed the outcomes all types of operations. However, in the midst of all of this activity, the environment itself remained static and unchanged. A bomb dropped on a truck may destroy the truck, but make no change to the underlying terrain or the surrounding vegetation.

Recently, this has been changing. Military simulation systems have included dynamic models of the interaction between military systems and environmental features. Simulated objects are able to knock down trees, crater roads, dig holes, build barriers, and destroy buildings. To support this, a new form of environmental model has evolved which understands the physical effects of vehicles and weapons on dirt, trees, and masonry block structures. Environmental modeling is no longer limited to static data structures, but includes dynamic models that respond to military operations.

DYNAMICS

To this point we have focused on defining and categorizing military modeling according to its application. Those categorizations were meant to illustrate the unique situations, problems, and interests of the developers and customers for military models and simulation systems. In this section we will describe the most dominant forms of dynamic modeling that are used in the community. Because military systems and problems are so diverse and such a large investment has been made in exploring them, there are many more unique forms of dynamic modeling than can be captured in a single chapter or an entire book. However, the forms that are described here are some of the most commonly used. They are also presented as general categories that cover a number of unique implementations.

Dynamic modeling is military simulation often focuses on activities like:

- Movement,
- Perception,
- Exchange,
- Engagement,
- Reasoning, and
- Dynamic Environment.

In this section we describe the dynamics that are included in each of these categories. This is followed by a section that explores multiple approaches to modeling these dynamics.

Movement

Dynamic representation of movement captures the change in an object's position over time. Models may represent position as a coordinate in two-space, three-space, or a velocity vector. Two-space coordinates usually include a position in X and Y, such as latitude and longitude. For models that represent only ground-based vehicles like trucks, tanks, and foot soldiers, this can be sufficient. The object may have no variation in elevation, or the elevation may come from the underlying elevation of the terrain on which it sits. Position may also include orientation, which in two-space would be limited to a 360 degree angle around the vehicle. A common reference system for this angle is with the zero point being aligned with true north and proceeding clockwise with 90 degrees being east, 180 being south, and 270 being west.

In three-space, the coordinate system includes a representation of elevation. This third dimension may be height above the local terrain, elevation above mean sea level, or distance from the center of a sphere that represents the Earth. The latter measurement evolved during the creation of distributed heterogeneous simulation systems. When networking multiple simulations, differences in the terrain representation within each system led to significant differences in vehicle position with respect to the terrain. Therefore, a non-terrain referenced coordinate system was needed to overcome these differences. When a three-space orientation is added to this model, it includes the pitch, roll, and yaw of the object, creating a six degree-of-freedom (6-DOF) model. When represented as a vector, this may also include the velocity of the vehicle along the axis of orientation.

In their basic form, movement models change these position and orientation coordinates according to a logical or physical representation of movement, as described in the next section. However, most implementations go further to include the effects that movement has on the object and the environment. The movement model may be linked to a model of the fuel consumption of the vehicle. This adds a limiting factor that can stop movement when the fuel is depleted. The inclusion of a fuel model leads to the need for the system to represent a process for replenishing the fuel consumed. Otherwise, the objects in the simulation will eventually grind to a halt. In military modeling, the addition of these details leads to the need for many more models to drive the additional variables that were added. Systems can grow far larger than can be developed, funded, or hosted on a computer through the poor management of these modeling details. Many authors have warned against this gradual creep in features that leads to the eventual failure of the system being developed (Law and Kelton, 1991). This type of growth is not limited to movement modeling, but can occur throughout the system if the designers do not control it.

A movement model may also calculate the number of hours of operation that the object has been used. This information is the root of most system failure and maintenance models. This drives a mean time between failure (MTBF), repair (MTBR), or other similar models.

The interaction of object movement with the terrain can generate environmental changes that trigger yet another model, such as the generation of smoke or dust clouds in the wake of a vehicle. If these changes to the environment are represented, then they call for specific environmental models that can calculate the size and density of the cloud created, as well as its drift and dispersion over time.

Perception

Military objects move about the environment in order to interact with other objects. One of the first steps in this interaction is to perceive or detect other objects. This is the process of applying a sensor to detect the existence, position, and identification of the other object. Sensor models capture the signatures of those objects. A visual sensor will capture reflected light from an object to the sensor. In most cases, the sensor model does not actually represent the path of a light vector, but instead considers the range and orientation between the target object and the sensor and calculates whether the target is potentially detectable based on the effective range and field-of-view of the sensor. A sensor model may also include information about the environment in which the detection is being attempted. For a visual sensor, atmospheric factors like the presence of smoke, dust, fog, and lighting may be used to diminish the possibility of detection. Also, environmental features like hills, trees, and buildings may be interposed between the target and the sensor and impact the detection of an object. The physical characteristics of the target may also be considered. Its size, contrast with the background, movement, and composition may significantly impact its detectability. Larger targets may be easier to see than smaller ones. Targets may have a higher or lower degree of camouflage, changing the ability of the sensor to separate them from the background image.

In military simulations, visual sensors are just one of a large variety of sensors that are available. Many systems include sensor models that collect signature information in the infrared spectrum, sound, emitted radio and radar signals, magnetic properties, and movement and vibrations. Models of each of these can be constructed at a number of different levels of detail, but each must determine whether to include the properties of the sensor, sensing platform, paired geometry, environment, target, and external interference. As illustrated earlier, as the sensor model becomes more complex, it drives the complexity of the entire system. Including all of the categories just listed would trigger the need for additional detail in the sensor model, but also the need for additional details in all target objects and the environment. Often the limitation in creating a high-fidelity sensor model is not driven by our understanding of the sensor, but, rather, by our ability to represent the characteristics of the target and environment that are needed to create such a model. In a military simulation system, the detail included in a model may be limited both by the needs of the customer and by the desire to keep the entire system balanced, not allowing one model to drive all of the others into become larger and more detailed (Pritsker, 1990).

Exchange

After moving and detecting, models are needed to allow objects to exchange materials and information with each other. Battlefield operations often lead to the depletion of materials like fuel, ammunition, food, medical supplies, and people. A logistics model may be used to represent the ability of the military to constantly deliver these materials to objects in operations. Such models are often based on an understanding of the rates of consumption, the pre-deployment of supplies to locations that are close to the operation, and the constant replenishment of supplies through a network of supply nodes. Replenishing supplies within n object on the battlefield is the culminating model of a much more complex representation of the logistics infrastructure that can stretch across an entire country or even around the world. The logistics model must also include mistakes and interference that cause it to breakdown and deprive the military objects of the supplies that keep them operating. A logistics model may be driven by textbook ratios of consumption or it may be include specific messages from the military objects about the levels of supplies consumed. In the latter case, a communications model is needed to carry information about what materials that are being consumed, by whom, and where they are located.

Communication is another model of exchange. The thing being exchanged is information rather than physical items. In the modern military, the amount of information that is carried around in a physical form, such as a book, letter, or paper map, is quite small compared to the amount that is transmitted in digital form. Therefore, modern models focus on communications in the form of digital computer and analog radio networks. A model of radio communications, like that of a sensor, may include the characteristics of the transmitter, transmitting platform, environment, the receiver, the geometry between the sender and receiver, and interference by other objects. Details in the representation of the radio or the signal it generates call for corresponding details in the receivers, environment, and countermeasures.

Military models of digital computer communications are similar to the tools used to study Internet traffic. They represent the senders, receivers, relay nodes, interference from competing traffic, multiple paths for the information to travel, and the loss of a message or the failure of a network. Modeling how people, objects, and units respond to the receipt of this information is included in the section on reasoning.

Engagement

Engagement is strictly a form of exchange. The item being exchanged is a weapon and the effect is the degradation of the operational capabilities of the target. Most military simulations perform movement, perception, and exchange specifically so they can put themselves in a position to engage an enemy target. Engagement has historically been the pivotal centerpiece of a simulation system and one of the most important models in the system. Certainly, not all objects engage the enemy, but those that do not are often referred to as support elements whose mission is to make engagement possible for combat equipped units (Smith, 2000).

An engagement model typically includes the exchange of weapons or firepower from a shooter to a target. This exchange decrements the capability of the shooter by expending ammunition in one of its many forms (e.g. bullets, missiles, bombs, rockets, grenades, artillery rounds). Just as in the perception and communication models described above, this exchange is usually impacted by the geometry between the shooter and the target. The engagement may also be mitigated by the environment and characteristics of the target. Trees, terrain, water, and buildings may interfere with the optimal delivery of the weapon and reduce its impact on the target. The target may also contain defensive systems that counter the effects of the engagement. A defensive model may represent the effects of flares or chaff in deceiving and misleading a guided missile or the protective effects of armor to deflect the weapon.

If the weapon successfully impacts the target and is powerful enough to overcome any interference or defenses, then a level of attrition must be calculated for the target. Different approaches to modeling attrition are described in the next section. Attrition is usually directed at the model state variables that control its ability to perform its primary functions. These may include health or strength, fuel levels, communications capabilities, and mobility. Models may also make a binary decision about whether a vehicle, human, or unit it completely destroyed or not.

The attrition model may be linked to communications and medical models. Communications models propagate the outcome of an engagement so that units or operators are aware that an engagement has occurred. These communications may trigger a medical model that will attempt to conduct extraction and provide medical treatment to simulated humans that are wounded. It may also trigger the logistics model to extract and repair vehicles.

Reasoning

Within large military simulation systems, there are usually many models of human decision-making and behaviors. These have become more prevalent as systems have grown in both the breadth of coverage and the depth of detail of the battlefields that are represented. Representing human thinking and even some computer reasoning are one of the most challenging parts of the current practice of military modeling. This type of information processing is largely not understood and general approximations and simplifications are captured in models.

Reasoning models often rely on the techniques developed within the Artificial Intelligence field. Techniques like finite state machines, expert systems, rule-based systems, case based reasoning, neural networks, fuzzy logic, means-ends analysis, and others are used to organize information and create decisions that are similar to those of living humans. FSM are currently the most widely used technique for both military models and those inside of commercial games. These reasoning models are challenged to perform a wide array of operations, to include commanding subordinate units, decomposing and acting on commands from higher level units, reacting to enemy attacks, selecting maneuver routes, identifying threats and opportunities for engagement, fusing information, and extracting meaning from intelligence reports. Each of these functions can be extremely complicated and require significant computing resources to execute. Reasoning models must balance their level of realism between robotic reactions to stimuli and detailed consideration of the situation prior to selecting an action.

The variety of reasoning models that are required on a battlefield cannot be fit to a single modeling technique. In practice, multiple techniques are required, each applied to a reasoning problem for which it is best suited.

Dynamic Environment

Earlier we described the evolution of the simulated environment from static state structures to dynamic representations of features and their interactions with military objects. Military objects interact with the environment both through direct intention and through accidental collocation. An engineering unit may be tasked to destroy a bridge or a road. This is an operation in which the effects on the environment are the specific intent of the action. In another case, an aircraft may bomb a convoy of trucks moving on a road. In this case, the trucks are the primary targets, but the road may sustain damage because of its collocation with the trucks.

Until recently, military simulations seldom included impacts on the environment. However, with the current focus on precision operations, there is much more interest in destroying specific buildings, roads, bridges, communications equipment, and pieces of the social infrastructure. Since this data is usually found in the environmental database, models that accurately modify environmental information are needed.

For decades, military organizations have worked on models that accurately represent the engagement that takes place between two tanks, airplanes, or ships. It is becoming necessary for those models to also impact the trees, terrain, and roads in the vicinity of these engagements. This means that information on the effects of weapons on trees is necessary, as well as their effects on buildings, roads, bridges, and a host of other types of surrounding terrain.

Even though these models are making the environment a dynamic part of the simulation system, the type and level of damage done to a tree is seldom the focus of the experiment or exercise that is being conducted. Therefore, the detail in these models is not as critical that in the models that govern the dynamic changes to other military objects.

MODELING APPROACH

In the previous section we discussed many patterns of relationships that exist between multiple models and talked in very general terms about what would be represented in those models. However, we did not explore specific mathematic or logical algorithms that would be used in those models. In practice, the number of techniques, algorithms, and equations that are used in military models is close to uncountable. It is not possible to describe all of them or even those that might be considered "the best". So many different problems are studied with military models that there is no "best" approach that can applied universally when representing a specific vehicle, human, or unit. However, the techniques that are used do fall into distinct categories. In this section we will discuss four categories of modeling dynamics that are often used in military simulation systems. Figure 2.1 illustrates this with a missile that can be modeled using any one of the four modeling categories to be described.

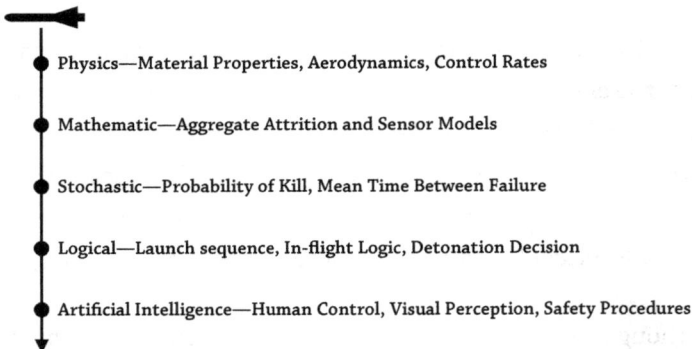

Physics—Material Properties, Aerodynamics, Control Rates

Mathematic—Aggregate Attrition and Sensor Models

Stochastic—Probability of Kill, Mean Time Between Failure

Logical—Launch sequence, In-flight Logic, Detonation Decision

Artificial Intelligence—Human Control, Visual Perception, Safety Procedures

Figure 2.1. Four different approaches to modeling the behavior of a missile.

Physics

Physics-based models are most often found in engineering and virtual simulation systems. For example, a missile pursuing a target would be represented by the physics of motion, momentum, mass, and aerodynamics. Changes in the fin positions would drive aerodynamic equations and change the vector of the missile based on the forces at work on the mass of the missile. Similarly, the seeker head in the missile would scan the environment electronically using the same pattern, revisit rates, and sampling rates of the real missile. This behavior would allow the simulated missile to collect data about a target in the same way that the real missile does.

Physics-based models are most often used to analyze the behavior of an existing weapon or to assist in the design of a new weapon. Understanding exactly how the pieces of the system will behave is an important part of exploring the design space to find optimum capabilities and combinations of capabilities that are optimum for the entire system.

Physics models require a great deal of data and mathematics. The data must be available for the system being modeled, the environment in which it is operating, and any other objects that it will interact with. Mathematics are required to represent a number of different behaviors of the system, interactions that occur within the system, and interactions that occur with other objects. Given this need, it is not sufficient to collect data and equations only for the missile that is to be studied. The model builders must do the same for the environment and for any objects that will interact with the missile.

Because of the volume of data, and the number and complexity of the equations that are required, physics models are necessarily reserved for smaller scenarios that involve only a few objects. Once constructed, the models can be computationally intensive. This means either purchasing a number of high-powered computers or accepting extremely long simulation times. The budget of the project limits the former and the schedule limits the latter. The project is literally a compromise of what the project can afford in time, money, and skilled staff. These limitations are one of the primary causes of the diversity in military modeling solutions. Constraints have forced generations of modelers to create unique representations of their problem.

Stochastic

Stochastic processes, probability and statistics, are most often found in virtual and constructive models. As simulation systems grow larger in their scope of representation, there is a need to capture many more activities and interactions in models. Lacking the detailed knowledge, breadth of expertise, access to data, time to build, and compute power to run a pure physics-based system, modelers have often resorted to a statistical representation of objects and interactions. In this case the models capture the behavior of many iterations of an event and represent individual event results using a probability function and the results of a pseudo random number generator. This type of modeling was introduced to the military modeling community by Stanislaw Ulam when he was working on the design of atomic weapons during World War II. Ulam encountered a number of problems for which the specific physical behaviors were not known, but where the pattern of outcomes had been measured. Therefore, he chose to use the statistical properties of the event and rely on multiple simulation runs to arrive at an accurate behavior for the entire system.

The previous missile example lends itself well to stochastic models. Instead of representing all of the minute physical interactions, a modeler could choose to represent the outcome of a missile engagement given a limited number of input variables governing each event and recourse to a probability distribution. The use of a pseudo random number in decision-making means that no one engagement contains all of the details of the event as in the physics model above. However, if the model is run a number of times, the randomness of multiple models will blend together and arrive at an accumulated result that is representative of the system behavior that emerges from all of the interacting models.

Stochastic modeling has proven to be extremely useful because it allows modelers to study problems that were previously beyond the limitations of physics models. This has led to the creation of very large simulation systems capable of representing hundreds or thousands of events and objects on a battlefield. However, these models also require that their creators understand both the physical behavior of the system and the statistical aggregation of those behaviors in order to create accurate stochastic models.

Logical Process

Physics and stochastic models are not appropriate for representing the processing of information that is carried out in a computer. These activities are better represented as a sequence of logical steps that make up a defined process. Within the missile there are controllers and computers that process information and make specific decisions based on those stimuli. A model of the missile may best serve the needs of a study by replicating that logic to control the missile's reaction to maneuvering targets or its response to control signals from an aircraft.

Logical models may also be used to capture the core rules of combat, or the steps that are followed by automated objects in carrying out their mission. These objects may be aircraft, ground vehicles, weapons, sensors, or any other battlefield object. When an object is controlled by a simulation system rather than a human operator, most of the time it is following a logical set of defined processes. These instructions tell it when to move, which direction to go, how fast to proceed, which objects to focus on, and which to ignore. These may be very complex processes, but they do not involve equations of physics or random decision points. In situations when an object should follow some form of "textbook" operation, the logical models are an excellent method of encoding this.

Finite State Machines (FSM) are often used to assist in organizing very complex sets of behaviors. FSM allow the modeler to capture hierarchical behaviors, triggers for changing from one behavior to another, encapsulated behaviors that can be reused in multiple FSM and deterministic behavior that can be mapped and validated. Military systems that are known as Computer Generated Forces (CGF) or Semi-Automated Forces (SAF) systems often contain a large number of FSM logic models. CGF systems are used to provide automated control of several dozen or hundred objects. A human may provide the overall mission and direction, but the CGF will supplement this with detailed control of movement and engagement through the use of FSM. These systems are not limited to logical models, but may integrate models of all the types described in this section. CGF have proven extremely useful in reducing the number of humans necessary to control simulated battlefield activities by moving detailed control from the hands of the human controller to the FSM logic.

Artificial Intelligence

Artificial intelligence also encompasses logic process models like FSM and production systems, but it is broader than that. In military modeling, these techniques are used to represent the behavior of humans, groups, and objects that are controlled by humans. The focus is on replicating the decisions that are made under a specific set of stimuli. To accomplish this, modelers and researchers have turned to FSM, expert systems, case-based reasoning, neural networks, means-ends analysis, constraint satisfaction, learning systems, and any other technique that shows promise in accurately capturing the complex reasoning process of humans.

The missile guidance and navigation example that we have been using is not ideal in this area. Though a missile model may use a FSM to model its behavior, it is not attempting to create an artificial representation of intelligence; rather it represents a logical process that is followed robotically. If the missile were being controlled remotely by a human who is viewing the target on a computer screen, then the behavior of the human might be represented using an AI technique. A neural network may represent the human's ability to discriminate a target in the scene and means-ends analysis may represent the human's decision process in selecting a target, leading its position, and switching from one target to another opportunistically.

AI techniques usually focus on processing information in a human-like manner. Using databases or rule sets, the algorithms attempt to make deductions that lead to behavior selection. The deductive process may be deterministic or stochastic (Russell and Norvig, 2000).

MILITARY SIMULATION SYSTEMS

Modeling is one part of creating a military simulation system. Within any one of these systems there can be a large number of models. Using the major categories of models described above, Figure 2.2 illustrates the relationships that often exist between these models to create a working simulation system. This figure includes only the major categories. For a specific system, the number of models would be much larger and the relationships between them would be more complex. This figure illustrates many of the causal relationships that were described in the earlier sections.

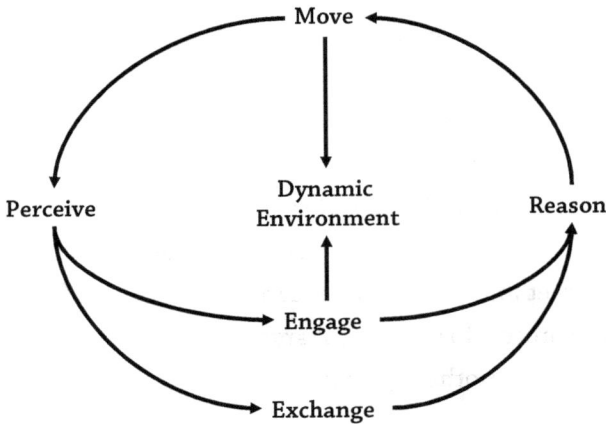

Figure 2.2. Model relationships for a simple battlefield simulation system.

The movement model is a good place to begin tracing the models in the figure. When this model calculates the new location, orientation, and velocity for an object, it may also trigger the dynamic environment model to represent the creation of smoke, dust clouds, or tracks in the sand. Once completed, the objects are in a position to execute perception models that can detect other objects from their newly achieved position. The perception model provides the necessary information to allow the objects to engage each other or to exchange information or material with each other. Engagement also triggers the dynamic environment models to create effects like road craters and destroyed buildings in the environment. An engagement between two objects may create collateral damage to surrounding trees, buildings, and roads. In some cases, the engagement is actually targeted at an environmental feature like a road or bridge. The exchange model calculates functions like refueling an aircraft or transmitting a message. Following the sequence of move-perceive-engage, the system may allow the objects to reason over what has just happened. This reasoning can take into account the results of each of the engagement, exchange, and perception, integrating them to enable the reasoning model to select the next action to be taken. Once completed, the cycle can begin again with a new objective that is received from a human user or from the reasoning models.

This cyclic diagram is a simplification of a real system. In actual implementation, the reasoning model may be activated at the completion of each of the other models, providing much finer over the decision-making process. That is characteristic of virtual-level simulations in which the reasoning component is providing very detailed control of a computer-controlled entity. The reasoning model may also be triggered much less frequently than the other models. This occurs in constructive-level simulations where the reasoning is at a much higher level of command and decisions are made infrequently with respect to the rate of activities in the other models.

CONCLUSION

This chapter has provided a high-level overview of modeling dynamics in military simulation systems. The very large number and variety of military systems that have been created, makes it impossible to describe the most common or "best" approach to modeling. Existing military systems focus heavily on movement, perception, and engagement. But, they may also include models of medical operations, communications, intelligence processing, military engineering, logistics networks, and command and control.

The mission of military organizations changes in response to the political situation in the world. The changes that have occurred in world politics are influencing the types of things that are modeled in military systems. Newer simulation systems are focusing more on communications, social influence, police actions, one-on-one interactions with noncombatants, and urban environments. These call for scenarios that study smaller interactions between military forces or between the military and the civilian populace, rather than large theater-level models involving thousands of combatants on each side.

Models of the threat or opposing forces are also changing significantly. New models are being created that represent suicide bombers, improvised explosive devices, riots and protests, and active avoidance of direct engagement.

The future of military modeling will include increasing level of dynamics in the modeled world. Rather than focusing only on the combat-relevant activities of an object, we will be creating objects that have a much more extreme range of dynamic properties. These "extreme dynamic" models will create a more realistic world in which the human users and the automated objects will be able to interact with the virtual world in all of the ways that a real person would. This could include being able to assemble primitive objects into more complex ones, breaking objects into multiple pieces, tapping into the electrical systems of buildings, digging holes in the terrain, or interfering with the normal operations of a vehicle by flattening its tires or inserting rocks in its gun barrel. Such a dynamic representation of the world is far beyond our current capabilities, due to limitations in both our modeling capabilities and the processing capacity of current computers. But, within a decade or two, military models will create a world that is "McGuiver-ready". This means that the modeling is so rich that a user will be able to do almost anything he can imagine in the world and a model will be there ready to represent those actions realistically.

Computer games like The Sims illustrate some of the richness that we are looking toward in military simulation systems. These games often focus on mundane activities like creating a meal, painting a house, mowing the grass, and reading a book. Though these activities will probably never be the primary focus of military simulations, they can play an important part in creating a realistic world in which to rehearse or experiment with military actions. During the Cold War, the primary military problem changed very little and this had a direct impact on the evolution of military models and simulations. In today's more chaotic and every changing environment, the military is being forced to look for ways to represent a much wider variety of objects and interactions. This will lead to significant changes in the dynamics that are modeled in future simulation systems.

Chapter 3

WAITING FOR THE DOT.SIM BOOM

S imulation has not had its dot.com boom. Our field has always grown in lockstep with a number of high technology areas. We have ridden right along with, and in some cases have driven, the leading edge of new technologies. As mainframe computers became workstations, and then PC's, simulation products leveraged all of this equipment and became better for it. As computer graphics moved out of the university research labs, we were quick to adapt these new technologies to our flight and driving simulators. We moved from large cabinet-sized image generators, to smaller graphics boxes, to dedicated workstations, to internal graphics cards. As networking and the Internet became prolific we connected our simulators together, created standard networking protocols, and constructed distributed events across all of our facilities.

We have always been very hungry for advanced technologies because our customers constantly demand more capability, better performance, and lower costs. Working in the simulation and simulator business has always given engineers the opportunity to apply the newest technologies emerging from research labs and commercial vendors.

But all of a sudden this stopped. The commercial world discovered that the Internet allowed them to do business in an entirely different way. They were able to connect directly to millions of customers around the world without creating physical stores and without shipping special equipment to every customer. The Internet opened the door to delivering products and services to every single customer in the world. With it the Amazon.com website could sell more books than Barnes & Noble with its 800 physical stores. This was a huge change in the relationship between a vendor and its customers. I grew up in a small town in Southeast Colorado with very limited access to retail products and professional services. A big shopping center was the Sears Catalog store. It contained a refrigerator, a dishwasher, and a table full of catalogs. If you needed appliances, lawn equipment, tools, or clothes you shopped for them in the catalog and placed your order. There were only a couple of storefronts on Main Street and not a single bookstore. My bookstore was a single small shelf at the local drugstore. My Amazon.com was the mail-in form on the back pages of the books I purchased from the drugstore. Barnes & Noble could not reach out to me with its vast selection. I had to reach into the inventory through the soda straw listed on one page of a paperback book.

Amazon.com is not just a huge warehouse of books. It is a delivery system that can reach every single person in the networked world. It allows a child in a small town of 500 people to access the same books as a child in the heart of New York City. It breaks down the location-specific barriers that prevent people from learning and exploring on their own initiative. That is the real power of the dot.com boom. In the simulation community we have not created this kind of service for our customers.

Simulation systems are still delivered like heavy products to specialized facilities. We create destination sites in the pattern of Disney World that soldiers have to visit physically to experience. And like Disney World, such visits can be once-in-a-lifetime experiences. Our soldiers cannot travel to our high-end destinations every time they want to improve their performance or explore a new idea. A dot.com boom in simulation would extend our systems through the military Internet to every soldier's desktop computer. It would allow every soldier to browse our offering of simulation services, enter the one of their choice, and join a team to explore a new idea or receive a lesson from a leader.

The technologies to do this are available now and pooling at our feet. But we continue to insist that training via simulation requires a dedicated facility, specialized equipment, and a large support staff. We insist that simulation cannot do a soldier any good unless it is custom crafted by an experienced professional and makes scant use of the newest technologies. To continue the Amazon.com analogy, we are insisting that books should only be sold in physical stores by a trained staff, and that an online bookstore would corrupt this process by allowing people to select their own books without explicit human guidance. This same argument was made against online education for years. Online universities were once considered the lowest form of crass commercialization of a much higher calling, not much above a diploma mill. But today, every university from Harvard to the local Community College offers some or all of their degree programs on the Internet.

The technical tools already exist to provide Internet-delivered, simulation-driven, training and exploration. What does not exist is the will to customize and extend our resources to reach every soldier in the service. We still want the soldiers to come to our specialized facilities and our dedicated staff. We are not ready to let soldiers take a hand in guiding their own training.

We currently have wargames that can be adapted to run on the server side of these training networks. We have computer games that can provide the intuitive client side interface. We have IT infrastructure tools that can tie the right soldiers to the right applications. New "Web 2.0" applications are opening the door to user created and modified content like our simulation scenario databases. These will allow a soldier to modify existing scenarios and select specific AAR products from within a standard browser. User interfaces like Google's Earth and Map products can provide a window into a simulation running in the military computer cloud. Though DOD security regulations may make the use of such tools difficult, they do not make it impossible. What is really difficult is to get our community to see simulation as a service that can be extended to millions of soldiers, rather than as a device, a facility, or a destination experience.

Each time our industry grasps new technologies there is the fear that we might abandon the good work that is being done with the older methods. But these fears are never realized. There was a time when all training was Live. The emergence of board wargames provided tools to supplement and extend that live training, but not to replace it. The computer revolution brought us virtual flight and vehicle simulators, but these did not replace live training. Instead they allowed pilots and vehicle crews to experience situations that were impossible to create in the live world. *Simulation as a Service* will expand and extend the value of our products to the soldier in the same way that virtual and constructive systems have done in the past. Some tasks can be trained very effectively from a standard desktop computer. Other skills must remain in the live, virtual, or constructive systems that we already have.

A dot.com boom in simulation will do the same for the soldier that Amazon.com did for the reading children of small town America. It will make huge volumes of training material accessible to every soldier, everywhere, all the time.

Chapter 4

SIMULATION AS AN IT SERVICE

"Innovation requires building a community of like-minded and wholly committed individuals who see their shared future in the success of the emerging technologies and industries" (Hargadon, 2003).

What are the emerging technologies and industries from which our community is drawing ideas for new types of simulation systems and services?

Within most organizations, including the military, mission-specific systems are developed, deployed, and maintained separately from business systems and training systems. In the military there is a clear distinction between the computers and networks that support real operations and those that are used to transact daily business activities or to train soldiers to do their jobs. The hardware resources are certainly kept separate, but each area also uses very different applications, purchases them from different vendors, and is conversant in different technologies. As a result, major improvements on one side of this three-sided IT-fence are seldom picked up and used on the other sides.

Over the last decade there have been major improvements in all information systems, to include hardware technologies, software products, infrastructure, and conceptual architectures. Huge commercial investments in business information technologies have driven that area from its small, back office orientation to the center of global business operations. However, the advances that have emerged in business IT are generally not recognized or adopted in mission or training systems. In this paper we will explore the viability of delivering training simulation applications to their users in the same way that IT services are delivered to business customers.

Separating mission systems from training and business systems was necessary in earlier decades. But the differences between these domains are being overwhelmed by similar complexities of hardware, infrastructure, software, and customer requirements. To discuss simulation as an IT service, we need to identify some common definitions for these terms.

IT: *"IT deals with the use of electronic computers and computer software to convert, store, protect, process, transmit, and retrieve information. For that reason, computer professionals are often called IT specialists/ consultants or Business Process Consultants."* (Wikipedia, 2007)

Service: *"A service is a well-defined, self-contained function fulfilling a particular business need provided by an application or module on request of another application."* (Strnadl, 2006)

Simulation: *"A system that represents activities and interactions over time. A simulation may be fully automated, or it may be interactive or interruptible."* (NSC, 1996)

IT SYSTEM CHALLENGES

Business IT systems are typically focused on the information necessary to allow the organization to function internally – e.g. accounting, regulatory compliance, human resources, finance, email, and the corporate web presence. Each of these was originally a different application and a separate infrastructure. But, many have merged into Enterprise-wide systems that interoperate in support of larger corporate functions. In supporting these functions, IT managers have identified a common set of challenges, demands, and limitations on their services. Some of these are:

- *Bandwidth*—always the first limitation in IT system service. Seeking adjusted data flow based on receiver bandwidth, timing, and applications;
- *Quality of Service*—requires dedicated bandwidth, shared availability is not sufficient;
- *Service Level Agreements*—defines the applications that the customer will have access to, response times that will be delivered, staff that will support these, and performance monitoring tools;
- *Access Validation*—controlling access to each application and data set;
- *Server Architecture*—use of data warehousing, middleware, and virtualization to connect data sources, software applications, functional services, and customers. (Williams, 2007)

A decade ago many of these terms and concerns were not priorities for mission-specific or training systems. The terms were seen as IT department specific and foreign to everyone outside of that department. However, as mission and training systems have become more sophisticated to serve globally dispersed users, provide different levels of access, and deliver numerous services simultaneously over a single network, these issues have clearly become important to the mission and training domains as well (Armour, 1999 and 2001; Boar, 1999; Zachman, 1987).

In this paper we introduce the idea of "Simulation as an IT Service". By this we mean a training system with the following properties:

- Professionally Managed
- Customer Oriented
- 24/7 Access
- Globally Accessible
- Facility, Geography, and Time Independent
- Light Clients
- Controlled Access

In the training arena customers need on-demand systems that can deliver services 24/7 and deliver these to facilities and computers around the globe. It is no longer sufficient to provide training services in a dedicated facility known as a "training center". These serve a single set of customers at a specified time according to the schedule of the training center. That is a 20th century perspective on training and IT. It assumes that the customer can be made to wait until the service provider has an available time and facility. Much of the training that we can deliver is actually a service, not a product and not a geographically-dependent experience. When this is the case, we can shuck off the limitations of our traditional mental model of training events and open ourselves up to the possibility of delivering services on the customers' terms. Meeting such a challenge requires investing time, money, and energy into information technologies that are beyond the capabilities of most legacy mission and training systems. To serve our 21st century customers the training systems community is going to have to adopt the techniques, technologies, and customer-focused mindsets that have been pioneered and implemented in the business IT domain.

MILITARY SIMULATION DEPLOYMENT ISSUES

Military simulations are typically delivered in the form of a stand-alone product that employs a unique hardware suite, tight client/server component connections, unique exposed data model and software services (e.g. an HLA FOM), and high demands on support staff. The mindset used in creating these systems is typically derived from the "big iron" perspective from large military systems:

- Heavyweight computer hardware,
- Dedicated computer networks,
- Tightly integrated software,
- Local support staff, and
- One-to-one relationships between the system (hardware and software) and the simulation event.

Given this approach, it is no surprise that many simulations are created, deployed, and operated in a manner very similar to military hardware systems like vehicles and weapons. Systems like Corps Battle Simulation, JANUS, and ModSAF typify this approach. These systems are very expensive to procure, deploy and operate. They place heavy demands on hosting facilities and the staff that support these. If a customer organization cannot support these demands, then they cannot use the simulation product. As long as simulations remain Local, Heavy, and Staff Intensive we will be limited in our ability to apply them to a diverse set of military training and experimentation problems. There are a number of domains which could use simulation if the usage costs could be trimmed significantly—something that "Simulation as an IT Service" is targeted at enabling. In many cases there is no need for every facility to host heavy software, expensive hardware, and large staffs when these could be accessed as a service from a remote simulation center. If we can break the one-to-one relationship between facilities, support staff, hardware suite, software products, data storage, and training event (Figure 4.1); then we can begin to deliver simulation-based training experiences to every military service person at every facility in the world—without delivering simulation-specific hardware.

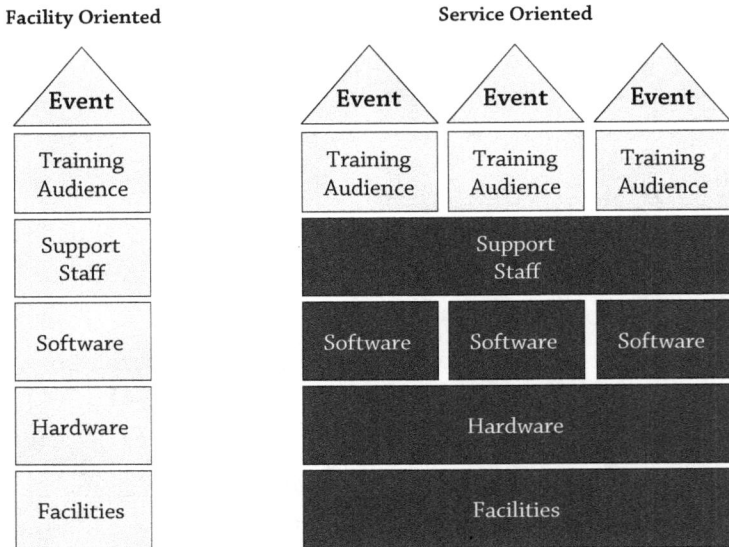

Figure 4.1. It is important to break the one-to-one relationship that exists from facilities through simulation events

EMERGING COMMERCIAL IT SERVICE EXAMPLES

"Web 2.0" has become a moniker for many of the new types of services that are offered over the Internet. In some cases these are truly unique applications; in others they are modifications of older ideas that have simply become accessible through the growth of network bandwidth, routers, and infrastructure. Services like Amazon Simple Storage Service (S3) and Elastic Compute Cloud (EC2), Google Earth and Maps, and Microsoft Live and Update have direct correlations to military simulation systems."

In a recent news magazine, the authors proclaimed, *"[Google's] new thrust represents a dawning era in corporate computing: software delivered as a service over the Internet, so it's accessible anywhere there is a Web browser handy"* (Hof, 2007). These types of statements have been floating around for years, but Google, Amazon, and Microsoft are finally beginning to bring them to the commercial marketplace. The concepts can also be applied to simulation as well – converting it from a product to a service.

Server-side IT Infrastructure

Amazon S3 and EC2 provide on-demand computational resources and storage which allows customers to purchase only as many resources as they need minute-to-minute (Amazon, 2007 and Yunis et al, 2007). Commercial companies are invited to scale their business IT infrastructure to meet the peaks and valleys of customer demand, rather than purchasing sufficient hardware to meet peak demands and allowing it to sit idle during lean times. Military simulation events have a similar operational dynamic. There are times at which we need significant resources for an experiment or exercise, but once the event is completed we can scale back to the minimal resources necessary to organize and package exercise data or to prepare scenarios for the next event. If our simulations could be designed to escape the need for dedicated machines and tight (local) client/server connections we would be able to apply this same approach to our hardware and facilities. We would no longer need numerous simulation centers stuffed with hardware that becomes outdated every year. Instead we could establish a military EC2 and S3 and use it as needed (Figure 4.2).

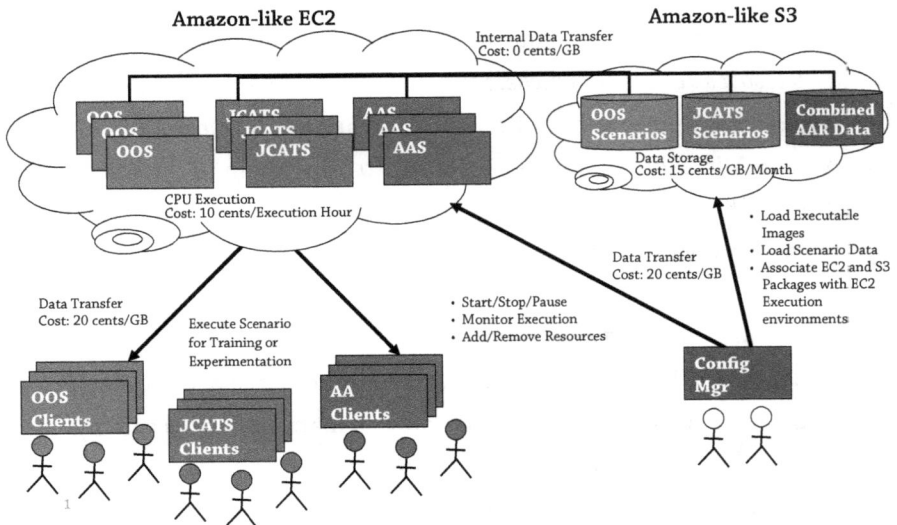

Figure 4.2. Amazon-like EC2 and S3 services use rented, network accessible hardware resources to host applications on demand

this model, the computing resources do not belong to a single facility, but reside in a globally accessible service center. They are rented and applied as needed by customers around the world and released when an event is finished.

Client-side Light Applications

To reach a global base of military customers, the client applications have to be designed to operate at a significant distance from the servers and to use protocols that can span open network topologies. Client applications of this type can be grouped into three major categories from very light to heavy (Table 4.1). The light applications may be simple HTML web pages or may include richer content such as Flash and Shockwave files. Medium weight clients may be similar to Google Earth or Java Web Start delivered programs. Heavy weight clients may be applications like the entire Americas Army or Second Life games.

Table 4.1. Three distinct categories of client-side applications.

Client Category	Examples	Conceptual Size
Light	HTML, Flash, Shockwave	Less than 2MB
Medium	Google Earth, Java via Web Start	2MB to 10MB
Heavy	Americas Army, Second Life	50MB to 100MB

Rasterwerks released a beta version of a 3D shooter game using Shockwave which demonstrated its viability for similar simulation applications. This game provides a single deathmatch level that is similar to those played in Unreal Tournament and Quake. But this game runs in a web browser and is downloaded as a Shockwave file in real time (Figure 4.3). The game is not installed on your local computer, but allows a customer to play from any consumer quality PC and network connection in the world. Phosphor also allows multiple players to interact in the world, includes simple AI-bots, high quality visual models, sound effects, and animated in-world objects. Its richness is equivalent to that provided by a 100MB installed game circa 2003—though the size of the area represented is more limited.

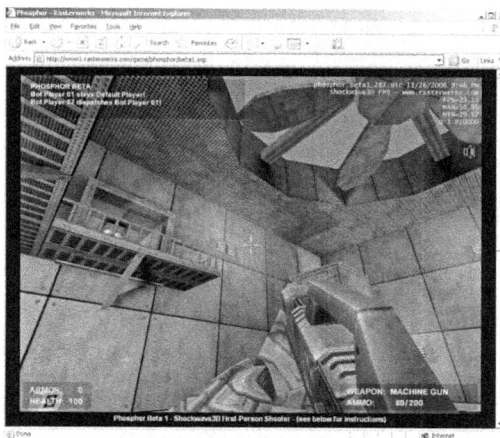

Figure 4.3. Web-based 3D shooter game that requires no pre-installation.
Source: http://www.rasterwerks.com/

Phosphor is a very useful approach to small simulation environments that require a limited amount of data and tie together a small number of players. Given a small client application that uses commercial protocols to connect to the larger server applications, a web-based client is a potential complement for large simulations. Systems like OneSAF and WARSIM could be engineered to operate in this way. The user interface for those systems is a two-dimensional plan-view display (PVD) that does not require sub-second map updates to be very usable.

A middle-weight client application would be something like Google Earth, as around 8 MB. Several companies have begun to experiment with Google Earth as the client map application for large simulations. In its native form Google Earth relies upon heavy servers for data and can connect to third-party servers to retrieve different layers of data to place on the map. Dynamic web page programming like Ajax enables web pages to update themselves to load the most current location and status of military units on a map. FboWeb has created a Google Earth application that performs similar functions in tracking the locations of airport traffic in real time (Figure 4.4).

Figure 4.4. Google Earth and fboWeb.com can track the locations of multiple aircraft in real time (1 minute update rate)
Source: http://www.gearthblog.com/

Given the aggressive growth of Web 2.0 applications and the financial and intellectual resources of companies like Google, we can confidently predict that future applications will be able to track large numbers of objects in real time—aircraft, traffic congestion, theme park customer movement, national logistics flow, etc. These capabilities have direct applications in military training and are one part of enabling simulation as an IT service.

Commercial Standards in the Infrastructure

The HLA community recognizes the importance of allowing a federation to communicate through the Web, similar to Web-based IT applications that dominate the business world. The creation of the Web Services Description Language (WSDL) API for the RTI opens a portal through which federations can deliver information to web-based clients, and hopefully suggests that this data stream should be designed to be light enough to operate on commercial Internet connections (Moller & Lof, 2006 and 2005).

This type of gateway may enable web-based access to simulations and be valuable for tools that are used in all types of domains—virtual, constructive, and live—training, testing, and experimentation.

CONCLUSION

There are a number of advantages to deploying military simulations in the same way that IT applications are deployed. We have explored a few of them here, to include:

- Reduced equipment ownership costs and obsolescence,
- On-demand user access to the best applications,
- Commercial architectures to access advances in IT practices,
- Centralized control of server applications,
- Currency of client applications,
- Interfaces between commercial and military infrastructure components.

"Simulation as an IT Service" refers to an idea that has a number of different implementations. The most important change is in what it allows us to do for our customers. Our community needs to see simulation as a service provided to a broad array of users, not as a hard product that is installed in geographically defined facilities. We need to provide services for training and experimentation that can be accessed by all potential customers around the world - subject to reasonable computer and network requirements. These services must not be limited by the physical location of the computer hardware and software application. Training services must not be limited by the number of professional staff required to operate the applications. We must break the one-to-one relationships that currently exist between: facilities, hardware suites, software applications, scenarios, support staff, training/experimentation events, and the training audience. Military simulation services need to be available to our users in the same way that Amazon, Google, and World of Warcraft provide access to their services when the customers need them, were the customer needs them, and independent of the number of other customers requesting service at the same time.

Games like World of Warcraft show that complex services can be delivered to thousands of simultaneous users and Amazon EC2 and S3 are enabling just-in-time access to the hardware necessary to do this. Together with the increasing performance of IT applications and architectures, wider network bandwidth, and more powerful desktops, this service-based approach to simulation delivery can become a reality within a few years. We need to get outside of the box we have lived in since board wargames moved onto computers in the 1970's. We need to explore new approaches to delivering training to customers—approaches that reach all customers and that are available 24/7. Gian Zaccai, the CEO of the Design Continuum, pointed out that, *"moving among industries frees you from the dogma of any one industry and their firm belief in the links between problems and solutions"* (Hargadon, 2003). In our community many people have a firm belief of what solutions have to look like. We need to look outside of our industry for innovative and different solutions to these problems.

Chapter 5

CLOUD SIMULATION

*S*imulation is currently perceived as an activity that is performed in conjunction with large amounts of specialized equipment. Over the last few decades, the military services have taken advantage of advances in electronics, computers, communications, and materials to create really outstanding training simulators, wargames, and the infrastructure necessary to connect these around the world. Soldiers are now able to training more effectively than ever before. The wide variety of venues and equipment allow us to focus training on specific capabilities to be learned and improved.

The success and power of modern technologies has made it possible to further extend the way simulations are created and delivered to Soldiers in the 21st century. Advances in computer chips now allow any citizen to own a machine with the power of supercomputer from the 1990's, and to carry that power in a laptop form anywhere they go. These same advances make it possible to deliver powerful training simulations in that same laptop, to connect to a global network of other Soldiers, and to access new forms of training from any common laptop computer.

In the first decade of the 21st century we ported a number of simulation applications to standard desktop computers running both Linux and Windows operating systems. This was a big step in reducing the hardware costs for simulation centers. But it still requires that the machine be configured specifically for simulation applications. In the second decade of this century we will have the ability to host the entire simulation application on networked server machines such that it can be accessed and run from within a standard web browser on a laptop computer. This simple change in paradigm eliminates the need for dedicated simulation client computers and allows every soldier to access any simulation using the standard computer issued by his IT department.

Moving the entire simulation application onto networked servers is often referred to as running it "in the cloud". This change will have three major impacts on training with simulations. First, it will significantly reduce the cost of computers necessary to provide training to the soldier. We no longer need dedicated client machines, but can leverage the existing investment in office IT machines that have already been deployed. Second, it allows a soldier or a unit to access training applications on their own schedule. They are no longer competing for access to scarce, dedicated simulation centers. Third, it eliminates all of the travel expenses associated with transporting units to specialize simulation centers. Fourth, it allows some training to take place while the soldier is at home station and is rebuilding relationships with his or her family.

As the military added computerized wargames and virtual reality-based simulators in the 1990's, there was a concern that these devices would attempt to eliminate the live training events that have been the core foundation of training for centuries. However, what actually occurred was the addition of new forms of training to cover skills and situations that could not be represented on a live range. The result was a complementary mixture of live, virtual, and constructive training that is more effective in preparing Soldiers for war than any training programs of the past.

Delivering simulations to Soldiers as a service on the network will extend this advantage and give us one more medium through which to teach Soldiers the necessary skills to be an effective warfighter. It will allow us to reorganize training so that it is more cost effective, more training effective, and more accessible to the soldier.

This new service will leverage a number of newly available computer technologies like—cloud computing, browser-based virtual worlds, secure networking, and user created content known as web 2.0.

WHAT IS TRAINING?

According to Dupuy, war became an art form when the Greeks invented the Phalanx. This new tactic could not be implemented ad hoc on the field of battle, but required a degree of coordination between the Soldiers that had to be perfected ahead of time in organized training sessions. This was one of the earliest forms of live training.

During World War I, Soldiers in boot camp were required to carve a "sham" rifle from a wooden two-by-four plank. All of the real rifles were being used in the war and could not be spared to train recruits. These early training devices allowed the soldier to drill and to practice, to learn many lessons that did not require a fully functional rifle.

In both cases, the training these Soldiers received was through their own actions, and only supported by the devices. Training often attempts to communicate new knowledge to a soldier's brain and the specific devices are simply the tools that allow this learning to happen. When a solder trains, it is an activity. When an experienced instructor leads training, it is a service. The devices that stand between the instructor with the knowledge and the soldier with the need are expedients. Historical precedent often leads us to embed the devices and facilities in our definitions of how training should be conducted. We forget that training is usually about transferring knowledge and skill, not about being exposed to the training devices.

Following the Vietnam War, the US Army created the National Training Center in the California desert. This 1,000 square mile facility has allowed the military to rehearse large battles with live equipment and live Soldiers for over 30 years. The center has been instrumented for data collection, combat adjudication, and live fire. Its ability to provide training is based on the open terrain, nearly unfettered movement, electronic instrumentation, and human staffing.

In the late 1980's the military developed the Simulator Networking (SIM-NET) project and devices. These brought the most advanced technologies in computing, networking, graphics, and control systems together to create a virtual training environment that could put multiple vehicle crews into the same virtual environment for the first time. It was hailed as a relatively low-cost training environment.

In the early years of the 21st century, the Army fielded the AMBUSH and VBS2 computer games to provide training in a "miniaturized" virtual environment. The graphics quality and control systems of these games are similar to those found on a much larger virtual simulator like SIMNET or the Close Combat Tactical Trainer (CTCTT), but the system makes no attempt to achieve physical and sensory immersion. These games can contain the same computational algorithms used in a virtual simulator. They are much more portable and can be distributed to many more Soldiers because of the lower cost of the hardware involved.

In each of these cases, the developers, the support staff, and the training audience can point to the hardware and say "that is the simulator." The perception has developed that training occurs with the device. But these devices are meant to facilitate training, not to **BE** training. Training is an event, a learning process, and a service provided.

Simulation as a Service will take training one more step toward universal access. It will overcome many of the limitations associated with expensive equipment at fixed sites. Delivering simulation as a service will allow Soldiers to access large numbers of training devices, but without requiring specialized and dedicated computer equipment.

Once deployed, this service will leverage the entire global network infrastructure so that the entire world is a circuit board into which any Soldier can plug-in from any location.

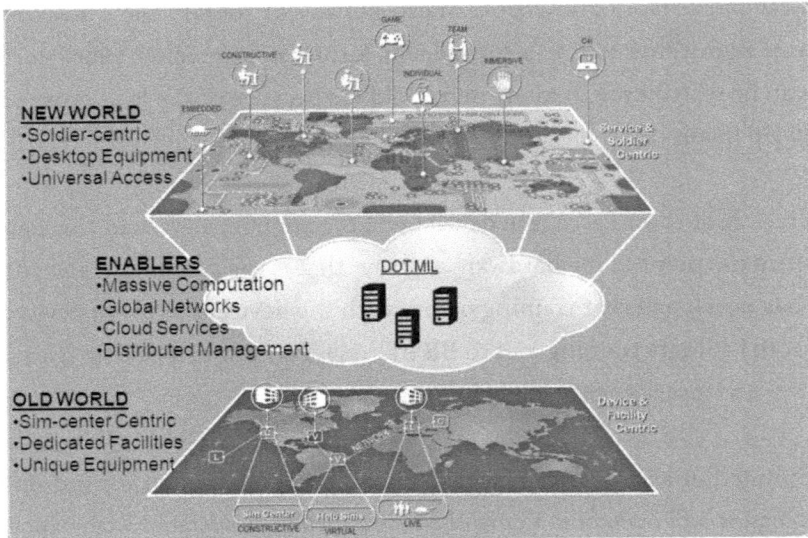

USING SIMULATION SERVICES

To illustrate the concept of simulation as a service we provide a number of use cases to show how it would work and the advantages that accrue from these services.

Terrain Familiarization

A group of Soldiers are gathered around a topographical map spread out on a table in the Battalion Tactical Operations Center. Led by the BN Operations Officer (S3), Major Monk, they are studying the terrain for their unit's advance tomorrow. They consider routes through and around small villages, across flat pastures and through rocky ravines. The map contour markings and their own annotations on their overlays identify the lay of the land, threats they are aware of, and the locations of previous engagements. As they do this, each of their brains in translating the information into an internal representation of what they will see and heard along the route. But each of them has a different picture in their mind.

These same maps exist in three dimensions in a number of mission planning systems. But that equipment is not part of their unit. They have several C4I computers and a few standard office laptop machines for word processing and web browsing.

The S3 brings up the Simulation Service web site in the browser of one of the laptops. He selects a specific geographic area, unit laydown, and threat database. All of these are loaded into the browser and he is looking at a 2D map very similar to the large paper version laid out on the table. But this map has validated combat models behind it. The icons on this map will advance and engage in response to commands sent back to the larger computers in the simulation service center. The S3 does not really know where that computer is and he does not care. All he needs is the web address on the dot.mil network and some familiarization with the tools for constructing a scenario.

He is able to create a 2D map of the environment and experiment with different approaches, all supported by the simulation service.

Using a 3D plug-in for the browser, Major Monk is also able to render this scene in 3D. The scene that is generated is very similar to popular computer games, though without some of the fancy shading and lighting. The other Soldiers in the TOC migrate from the paper map and stand around the S3 watching and directing his actions. But he tells them to sit down at the other computers and join him in the virtual environment. They can tag along over his virtual shoulder. Or they can explore their own ideas in the shared 2D and 3D space.

Sitting shoulder-to-shoulder the staff collaborates both physically and virtually. They switch between the paper map, the white board, and the computer simulation space in the computer. They alternate between working on their own machines and clustering around one person who is doing the driving.

Major Monk was introduced to this simulation service at the Infantry School where it was used for classroom instruction and instructor-led exploration of the battlefield. But it is a service on the Army network that is available to all Soldiers. The classroom tool has become a mission planning tool. It will also be used to review what happened after the unit has completed the mission. And all of this data will populate a shared database. With proper clearances, officers and non-commissioned officers who are still in the Infantry School will be able to access this database tomorrow. But so will other units all over the world.

Single Player Wargame

Wargames have been added to the Maneuver Captains Career Course curriculum. The big, old, clunky versions have been around for years, but were generally used to support Brigade and above level exercises and required a select group of individuals trained in their use. These new ones live on the military Internet and any student can load them up one any computer. Captain Daconta points his web browser to the wargame's web site and starts hunting for the prebuilt scenario that was assigned in this morning's class. The mission is a company raid on a suspected arms cache in a small village. To accomplish this, he will need to optimize the use of his Soldiers and equipment, plan the best avenue of approach, establish a cordon around the objective, conduct actions on the objective successfully, and save the results for grading.

After selecting the scenario, the browser pops open a Java window and begins to load the program and the data into a sandbox on his computer. Since the program does not need to install on his machine, he does not have to get system administrator privileges or move to a special machine where the simulation was previously installed by the IT department.

On the 2D map he lays out his company. Then he selects primary and alternate routes and plans initial reconnaissance utilizing a UAV. He enters his primary route and begins the simulation. As the commander of this operation he is watching the threat map and the simulated blue force tracker data. These tell him what is happening to the platoon he has out front. The UAV reveals an enemy sniper on a roof and he directs his platoon leader to the threat, where the platoon leader eliminates it before the remainder of the company advances The simulation navigates the company across some rough terrain and through a sparsely populated market area. They arrive at the objective area, a compound of a suspected arms merchant. Daconta notices that it has taken the unit a little longer to get there than he had expected, and probably longer than will get him an 'A' on the assignment. He scrolls back in time and looks for the major hang-up. He chose to skirt the edge of town rather than risk operational security by pushing his convoy through the early morning traffic in the heart of the village. He thinks that this route is a better alternative than the slightly faster, but more crowded, path he could have taken.

He lets the scenario proceed as planned. Once the cordon is in place, the simulated company launches several micro UAV's, just as he had programmed, to get a look at the back, top, and inner courtyard of the compound before making entry into the compound. With this information they are able to move into the courtyard and suppress any opposition with minimal shooting. The arms cache is in the cellar as reported and they have secured it in good time with no casualties.

But is this the best outcome? Will his classmates find a better solution? Captain Daconta restarts the simulation and works up a new plan to see if he can do better. He is determined to work this all night if that is what it takes to get the highest grade in the class.

Team Tactics

Master Sergeant Lanham has just received mission orders to conduct a cordon and search on a multi-level building that is suspected of housing a drug lab. The mission will happen tomorrow morning and his unit has 12 hours to prep and rehearse. They have worked on urban operations all week, such as movement in stairwells, breaching, hallways, and room clearing at the shoot-house. But now they want to get more specific with data on the actual building.

He gathers the detachment in the operations center with their standard issue laptop computers. They set-up a network connection and pursue intelligence data on the area of town where the building is located. Finding recent photos and a schematic, they tag that data as a scenario for their online Thunder simulation. Lanham visits the Thunder.mil web site, loads the intel data and prepares a force laydown for his unit. He includes the last kilometer of approach to the target building and the breach points for his men. Then he directs the rest of the team to load the Lanham-Thunder scenario he has created. They all find themselves on a 2D map of the area in their specified starting positions. MSG Lanham briefs each of them on the plan and they run through the scenario as a team in 2D. They provide feedback and Lanham adjusts the plan. They run through it four more times in an hour.

The 2D mission execution is then transferred to a 3D world. Lanham and his team enter in 3D and just watch it play out. They notice a few weaknesses in the plan and adjust. Finally, they take control of the characters in 3D and fight the plan as they have rehearsed. They become familiar with the appearance of the building, prominent landmarks, concealed positions, and the dangerous corners inside the building.

After an eight hour session they have learned all they can in the simulation service. They save the final scenario and mission plan. This will be the basis of the AAR once they have completed the real mission and are ready to review and capture lessons learned.

Leadership View

Lieutenant Colonel Wheeler has assigned his infantry battalion a number of scenarios to rehearse over the next week. They have been using the range and their instrumentation to practice it live. He has walked them through it multiple times on a paper map. But today they are all playing it out in a simulation. They are all at their desks using their standard office computers, but today they are using the simulation service, rather than working on presentation slides. Each of them has his head down playing a role in the game. He has dismounted and mounted infantry, their armored vehicles, and UAVs working.

Wheeler is watching from "On High". That is what the Soldiers call the commanders portal that allows him to view the battlefield from 1,000 feet in the air, from anyplace on the ground, or through the virtual eyes of any one of the Soldiers. Many commanders run two or three or four of these portals so they can see what's happening from many perspectives at the same time. If they are fast enough they seem to be everywhere simultaneously—that is Wheeler.

They are all chattering on the virtual radios that are simulated by the game, delivered by VOIP, and jacked into the standard Xbox headsets they are using. Wheeler pops in on various networks to issue commands, provide advice, and ask for updates. Once they are happy with their plans, Wheeler almost always has them change sides and take up sniper positions as enemy soldiers in locations of their own choosing, which often resulted in some significant changes.

The scenario runs for 90 minutes, but it seems like hours have passed when they come out. They have run through it from different angles four times and have gone through small critical pieces a dozen times. Everyone from the newest private to the commander feels like they understand this patch of the city in more detail than they have seen while patrolling it. The simulation data is stored on the server for them to use again in the future. The event is also cataloged in their online training record for credit.

EVOLUTION OF THE TRAINING ENVIRONMENT

Live simulation is a very physical experience. Soldiers get out on a controlled battlefield, they use their real equipment, and they go through as many real actions as possible. In Live simulation Soldiers fly their real helicopters, carry their real rifles, and have to work on the real terrain and in the real weather. This is generally the gold standard in training.

But this kind of training requires a lot of planning, coordination, and money. Flying a helicopter is an expensive process. Shooting live tank rounds is dangerous. And rehearsing the movement of thousands of Soldiers takes a great deal of space. Virtual simulation allows us to move some of these activities into a 3D environment with simulated vehicles and soldier avatars. This eliminates most of the danger associated with real weapons and significantly reduces the cost of the training. It is not exactly the same as practicing with the real equipment, but a soldier is allowed to repetitively take actions that are dangerous and prohibited for safety or environmental protection reasons in live training.

Constructive simulation is the use of computerized wargames that represent large areas and large numbers of people and equipment. These construc-

tive simulations have to have enough internal intelligence to be able to partially automate the behavior of hundreds of computer controlled objects. These
act as an advanced form of chess board that unit commanders and their staffs
can us to rehearse the movements of thousands of people and vehicles on a
simulated map. These focus on good tactics, planning, and timing on the part
of commanders who have to orchestrate the war.

Modern 3D computer games sit on the cusp between virtual and constructive. They provide a rich, 3D, immersive environment. But the soldier is interacting with the simulation via a desktop computer that is similar to those used
for constructive simulation. This kind of tool is very useful for teaching tactics
to lower level unit leaders who are walking the streets with their Soldiers and
need to learn the best methods of approaching situations in that environment.
They are developing their thinking, planning, and communicating skills.

Delivering a simulation to any computer as a service moves the advantages of constructive and game-based training to every soldier's computer. It
is as universally accessible as web-based email or online document editing. It
removes the constraint of being tied to a uniquely configured computer. Potentially, any military computer with a network connection can reach out and
pull in a wide variety of simulation-based training materials. These might be in
the form of constructive, map-based simulations or as 3D game environments.
But the key is that Soldiers will not have to install them on the computer, instead they will access and use in the same way that they access any other web-
based service of web site. This could turn every computer into an on-demand,
custom simulator.

Live, Virtual, and Constructive simulation systems are becoming part of
a more integrated LVC environment which draws on the strength of each of
these domains. Very soon, LVC will be joined by "G" as games are tied into

these integrated environments in a manner similar to many of the current virtual systems. Once simulation as a service is deployed, it will both serve as an infrastructure for tying LVC-G together, and as a new member of the integrated federation. Soldiers will join training events from their laptop computers and web browsers and play a role in a larger scenario. This will add the service component to create an LVC-G-S environment.

Acknowledgement
Many thanks to SGM Dave Lanham for his assistance in describing the military scenarios in this chapter.

Chapter 6

SYNCHRONIZING DISTRIBUTED VIRTUAL WORLDS

Distributed virtual worlds are a specific implementation of simulation, graphics, and networking technologies that have been evolving for decades. It is only recently that each of these fields has become sufficiently advanced to support consumer products like networked computer games and multi-user web-sites. As these applications have grown in popularity, the developers have been searching for techniques to keep the distributed pictures of the virtual world synchronized. This is necessary to insure that each player or user is experiencing the same version of the world, with the same cause-and-effect relationships between observed phenomena.

Networked Multi-Player Gaming Example. Computer games are an excellent way to illustrate the problems that arise when distributed users interact with each other. In a typical 3D-shooter game, a player enters a world model that is populated by monsters and other interactive players. Monsters are usually controlled by software residing on each player's local computer.

Synchronizing the monster's actions with those of the player shooting at them is relatively easy and is not the topic of this article. However, it is much more challenging to synchronize information about all of the distributed players on computers around the world.

While running through a dungeon, your location and position (actually, the location and position of the graphic avatar that represents you in the game) are calculated by the simulation algorithms loaded on your local computer. That information is then transmitted via the network to all of the other players in the dungeon. Of course, it takes a small amount of time for that information to travel to all of those remote computers and for those computers to read the information in and change the position of your avatar. Imagine that your avatar is hiding behind a large crate and you order it to peek out from behind the crate and duck back immediately. The network message that moves your avatar out from behind the crate may travel to the other players very quickly. But the following message, the command to duck back, may be delayed along the thousands of miles of computer cable that make up the internet. This delay may be caused by variations in the processing load at any point along the way, increases in network traffic, failed computer equipment, or many other factors. And, because other players are often spread all over the world, the "duck back message" is usually delayed a different amount of time in reaching each of the other players' computers. In this situation, the avatar on your local computer may have only been exposed for 1 or 2 seconds. However, the avatar on all of the other computers will be exposed for a much longer period. It will be frozen in the exposed position because the "duck back message" has not arrived on those computers. Therefore, when your opponents look around the dungeon, they see your avatar standing frozen in plain view. Of course they unload multiple rounds from their plasma weapon into your character.

You were robbed—killed by a network lag. However, interestingly enough, when the plasma bolt from a remote shooter arrives at your simulation node, it informs you that you have been hit while standing in plain view, but that message arrives long after you have ducked back behind the crate. Your simulation algorithm has little recourse but to destroy your character even though you are safely hidden. Failure to do so would create an inconsistency from the perspective of all the other players who see your avatar standing at the impact point of the plasma bolt.

Parallel and Distributed Simulation Technology. The example above illustrates one of the fundamental issues that arise when developing distributed virtual worlds or simulations that interact over a network. The delay imposed by the computer network interferes with the true pace of events and can corrupt the cause-and-effect relationships between these events. The problem is relatively new to the developers of consumer products like computer games and web-sites, but it has been around for several decades and scientists in universities and industry have been creating methods for addressing it. This type of research was initially focused on a class of applications known as discrete event simulations (DES), which are often models of factory production, seismic data analysis, laminar airflow studies, and visualizations of nuclear explosions. The techniques developed for those problems have been applied to distributed military wargames as well. Simulations of combat forces belonging to each military service were the first to implement some of the more advanced synchronization techniques in an interactive environment.

In this article, we will describe the leading techniques for synchronizing distributed virtual worlds and will provide examples of the algorithms that accomplish this. Readers who require a deeper understanding of any one of these techniques should consult the references listed at the end of the article.

DISCRETE EVENT SIMULATION

Since distributed synchronization techniques began in the discrete event simulation (DES) field, we should begin by describing the basic concepts behind DES and showing the similarity to virtual worlds that are just now emerging. A DES attempts to capture information about the real world in a form that can be used to study or illustrate its dynamic behavior. A photograph or 3D model of a factory is a static picture of that system. But a DES of a factory can capture the behavior of each machine and the cause-and-effect relationships between the machines when they are running.

Fundamental Components

The fundamental pieces of a DES model are State, Events, Transition Functions, and a Time Advance mechanism.

State. The state of the system is simply a group of variables that describe the system at a specific point in time. In many ways the state is equivalent to a static photograph of the system. It captures all of the important variables, but can represent only one or a few values of these at any one time. The state variables of one avatar in a computer game may include its position, orientation, rate of movement, weapon list, ammo count, and health.

Events. Events are interactions that occur within the system that cause it to change the value of one or more state variables. These events enter the system from some external source. For a DES performing an analysis of a factory, events are often pre-loaded in the initialization data and read into the simulation when it is started. In an interactive environment like a game, events enter the system as a result of a player entering orders via a keyboard, joystick, or other control device.They can also be generated by automated behaviors of computer-controlled monsters in the game. Examples of events that change the state variables of a game avatar include Move, Duck, Pick-up Weapon, Fire Weapon, and Explosion.

Transition Functions. Events arriving in the simulation do not have a magic power to locate state variables and change them. The association between specific events and specific state variables is made by a set of transition functions. These are the dynamic modeling part of the simulation. A transition function may be contrived specifically to handle a one type of event. It understands the format and content of that event and the relationships to specific state variables. This transition function calculates the type or magnitude of change that an event has on each state variable. A transition function known as "Movement" may be responsible for reading in a "Move" event, calculating its effects, and changing the position, orientation, and rate of movement of an avatar in a dungeon.

Time Advance. Events being processed in a simulation or virtual world must be sequenced according to some criteria. In most cases, this is done by the time at which the event was scheduled to execute. Therefore, some time advance mechanism must exist to move simulation time forward to allow future events to be executed. This is essential when ordering all the events from many remote sites. Because every simulation does not necessarily want to advance time in the same manner, there are several different types of time advance mechanisms.

Event Processing

During execution a simulation may generate hundreds or thousands of events. These queue-up to be executed in the appropriate order. In an interactive computer game, like a 3D shooter or a real-time strategy game, the computer may be receiving events from two or two dozen other players on the network. The software must attempt to order those events into a single list that is in the correct causal order. Causal order means that if a monster throws a fire ball at the player you are looking at, your simulation processes the movement of the fireball before it processes the destruction of that other player's avatar. It would be causally incorrect for the avatar to burst into flame before the fireball was even thrown.

Time-Stepped Simulation. For simulations that contain a large number of objects that are constantly changing state, it is natural to treat simulation time advance like the advance of a clock. The simulation clock ticks like a discrete watch and every object is updated to represent its state at the new time. Each game avatar moves to a new position, fires a weapon, takes damage from incoming weapons, etc. In a 3D shooter game these time-steps must be very small so the player does not see the virtual world jump from one position to the next. This is accomplished in the same manner that a movie is played. When a movie is run at 30 frames per second, the human brain perceives the series of pictures as a constant motion.

```
debtTime = 0;
while (!done) {
    startTime = getClock;

    <Do simulation processing here>

    elapsedTime = getClock - startTime;
    sleepPeriod = (stepSize/stepRate) - (elapsedTime + debtTime);
    if (sleepPeriod <= 0) {
    debtTime = abs(sleepPeriod);
    }
    else {
    debtTime = 0;
    sleepFor(sleepPeriod);
    }
}
```

Event-Stepped Simulation. In some simulations events are so few and far between that time-stepping results in long periods of processing where nothing happens. Material in a factory may flow from one machine to the next every 20 minutes. Modeling this at 30 frames-per-second would be a huge waste of processing time and power. It would be better to queue each event and process them in causal order, but allow the time stamp on the event to set the simulation clock time. Simulation time would then jump from one event time-stamp to the next without representing all of the values in between. This can save a huge amount of processing and allow the simulation to finish in a much shorter period of time.

Event-stepping a simulation can also be used in interactive simulations with many events occurring simultaneously. When the simulation is structured this way, an additional mechanism must be implemented to notify objects when other objects have done something that is interesting to them. In a game, a remote avatar may run through the room your avatar is standing in. But unless your avatar has scheduled some form of event at that time, an event-stepped simulation will not pass thread of control to your avatar, and you will never see the other player run by or have a chance to react to him. To capture thread of control in this situation, your avatar must register some form of interest in objects that come near you. When this happens, the infrastructure managing the events will alert your avatar to the presence of the other player, giving you an opportunity to react to him. Though this mechanism works, it is rather cumbersome and can still miss notifying you of events that you are interested in. That is why most interactive simulations are structured as time-stepped simulations.

```
while (!done) {
    getNextEvent(smallest time stamped message in the queue);
    simTime = eventTimeStamp;

    <Process this event. Send it to the appropriate object or event handler.>
}
```

Parallel and Distributed Simulation (PADS)

Like computer games, DES can be designed to executed on parallel computers or a network of distributed computers. These systems need to be synchronized for many of the same reasons that computer games must be synchronized, though the visual picture is seldom as easy to paint.

The PADS community has a rich history of inventing techniques to address the synchronization problem. For the last twenty years they have been solving problems that are only now beginning to emerge in consumer products.

DISTRIBUTED VIRTUAL WORLDS

With the advances in computer simulation, graphics, and networking, technologies it is now possible to create distributed virtual worlds for the general consumer. These began as advanced experiments for military applications. Flight simulators were developed to train pilots in essential aircraft control and tactics for engaging enemy targets. These simulators were then linked together to allow two or more pilots to fly in tandem or to compete against each other in realistic, non-lethal training environments. But these connections were usually limited to a few simulators using a communications protocol that was unique to the operating environment of that system.

Networked Military Simulation

In 1983 the Defense Advanced Research Projects Agency began the Simulator Networking (SIMNET) project in which they created an economical way to link a family of tank simulators to allow multiple crews to train in the same virtual environment. The principles learned under that program became the seeds for much larger distributed interactive simulation technologies that emerged in the ensuing years.

Tank Simulator Computer Game
Courtesy of MaK Technologies, Interactive Magic, and Zombie Studios

The techniques were also applied to the wargaming community where they were used to join wargame simulations for military staff training. These wargames were the first to implement time synchronization techniques developed under PADS research projects.

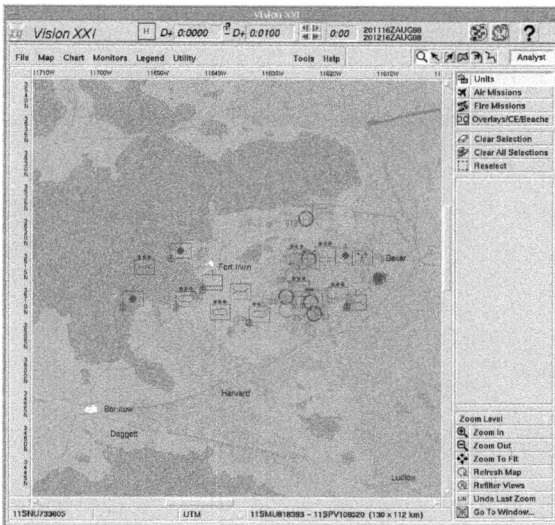

Military Wargame Map
Courtesy of Tapestry Solutions Inc

PART 1: MILITARY SIMULATION TECHNIQUES

Computer Games and Web-sites

Today consumers are eager to play interactive 2D and 3D simulations of all forms of combat. Therefore, as the PC has evolved into a truly powerful computing platform, the game developers have adapted military simulation ideas into new types of games. It is now possible to purchase a game form of every type of military simulation that was previously available only on powerful workstations and specialized image generators. In many cases, the visual and modeling capabilities of these games exceeds that of the best military systems.

In the past few years a huge market has developed for networked games that allow players to compete with each other from all over the world. These games face the same synchronization problems that have challenged parallel research projects and military applications for two decades.

As the World Wide Web continues to grow in sophistication, it will also evolve sophisticated synchronization requirements just as networked games have. Techniques for solving these problems already exist. In the following sections we describe the dominant methods that have emerged for synchronizing virtual worlds. Each of these can be applied to specific problems, but no single method is best for every problems.

INDEPENDENT OPERATIONS

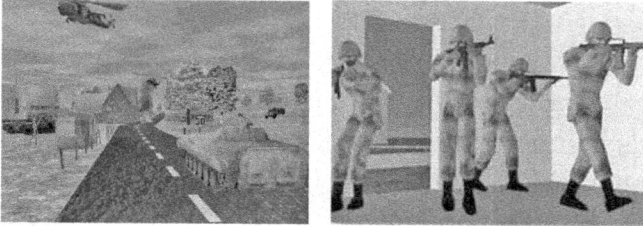

Linking Diverse Military Training Simulators
Courtesy of Evans & Sutherland and Boston Dynamics

It is **not** absolutely necessary to implement a synchronizing protocol between distributed virtual worlds or simulations. In fact, many commercial and government products exist in which the events are processed independent from the processing being conducted on other computers. These have no concept of synchronization.

When allowing distributed applications to process independently, each event published onto the network arrives at the other simulators and is processed in a first-in-first-out order. This approach relies on the computer network, processing hardware, and operating system to deliver events in an efficient manner with negligible delays.

Slight improvements can be made by time-stamping every event and using that information to order events in the processing queues as they arrive. Since events are being processed FIFO and immediately, the advantage is only realized when events arrive faster than they can be processed.

Independent processing of events is usually implemented using the UDP network protocol. This eliminates the network traffic associated with acknowledging message receipt and calling for retransmission when a message is lost or corrupted. It makes more network bandwidth available for the transmission of events and object attribute updates.

Since each simulator is independent of all the others, there is no way to identify the difference between a simulator that is processing slowly and one that is having system problems that prevent it from updating its objects. An avatar in the virtual world may freeze in place for an inordinate amount of time. This could be caused by slow processing at the sender, network congestion, system failure, or a number of other issues. In the case of a system failure, the avatars from that simulator should be removed from everyone's virtual world. Otherwise the objects are not changed in response to explosions and other events that impact that object. To minimize this type of problem, the defense simulator community has implemented the concept of a "heartbeat". Each simulator is required to publish a state update message every five seconds. Since virtual world updates are usually much more frequent than that, in the neighborhood of 15 per second, even a slow processing simulator can achieve this easily. We can then deduce that avatars that are not updated within the five-second period must have experienced a system failure and should be removed from the virtual world.

In virtual simulators, the images on the screen typically update at least every second. This, along with the mandatory heartbeats, can create a very heavy traffic load on the network. To reduce this most simulators have implemented some form of dead reckoning (DR) on object movement. Given the last position, orientation, and velocity of a vehicle a remote simulator can extrapolate the position of that vehicle as long as none of those attributes change. Then it is possible for active objects to send much fewer messages about changes to their position. Dead reckoning can be done is many different ways, but three of the most common are given in the table below. Zeroth-order DR simply assumes that a vehicle remains in its previous position until told otherwise. First-order

DR extrapolates the position of the vehicle based on its last known velocity and the elapsed time since the position was given. Second-order DR includes the acceleration rate of the vehicle in the extrapolation calculation.

Example Algorithm: DR equations

Zeroth-order DR:	DrLocation = lastKnownLocation;
First-order DR:	DrLocation = lastKnownLocation + lastKnownVelocity*timeElapsed;
Second-order DR:	DrLocation = lastKnownLocation + lastKnownVelocity*timeElapsed + (1/2)*lastKnownAcceleration*(timeElapsed)**2;

The independent operations described in this section were developed by the military simulator community and largely defined under the SIMNET and Distributed Interactive Simulation (DIS) programs. More details on the implementation of these can be found on the web site of the Simulation Interoperability Standards Organization (SISO)—http://www.sisostds.org/.

CONSERVATIVE SYNCHRONIZATION

In some applications it is essential that the distributed pieces of the system be strictly synchronized to allow them to vary the rate of progression of time, achieve identical event ordering, and stop or restart the simulations in a coordinated manner. These types of simulations often require regularly saving the state of the simulation and restoring that state in a synchronized manner. Wargames are also able to progress forward at different rates, such as twice the speed of real-time or half real-time.

"Conservative synchronization" is an umbrella term for techniques that keep all of the processes causally linked at all times. It may be implemented with many different structures, to include:

- Master Clock,
- Token Passing,
- Client/Server,
- Chandy/Misra/Bryant Algorithm, and
- Aggregate Level Simulation Protocol.

Master Clock. Synchronization can be accomplished by assigning one of the distributed simulations as the owner and controller of the clock. That simulation processes its own events and moves the clock forward, sending its time to all of the other simulations before it processes its own events. Those simulations accept that time and process events accordingly. Events that arrive from around the network are held in a queue until the master clock indicates that they can be processed.

Simulations that are slaved to the master clock are expected to operate fast enough to remain apace of the clock owner. As long as the slaved simulations run at least as fast as the master this approach works fine. However, if the shared simulation time must be mediated by all members of the virtual world, a more advanced mechanism is needed to keep the systems synchronized.

Example Algorithm: Master Clock Synchronization

```
Clock Owner:
while (!done) {
    simTime = incrementTime(prevSimTime);
    sendTime(simTime);

    startTime = getClock;

    <Do simulation processing here>

    elapsedTime = getClock - startTime;
    sleepPeriod = (stepSize/stepRate) - (elapsedTime + debtTime);
    if (sleepPeriod <= 0) {
    debtTime = abs(sleepPeriod);
    }
    else {
        debtTime = 0;
        sleepFor(sleepPeriod);
    }
}
Clock Slaves:
while (!done) {
    simTime = waitTimeUpdate();

    <Process events at this sim time.>
}
```

Token Passing. In some situations, permission to publish and process events can be controlled by exchanging a network token. As an example, synchronization in an online, four-hand, bridge game can be controlled by passing a token from one player to another. Information about each card played is published to all of the other players allowing them to see what is played. But no player is allowed to play a card until they receive the network token that is cycling between all the players.

This method is very useful for turn-based games in which a fixed number of players join a game and remain in the game from beginning to end. A game server usually exists to match players together to begin the game, but the server does not have to play an active role in the progression of the game. If the game is one in which players will join and leave as it executes a server must be involved to accommodate this. It requires that the server receive control of the token at the end of a cycle of play or between each player's turn.

Most turn-based card and strategy games use a server to keep statistics and ranking boards on each player. The server contributes a certain amount of community to the game by preservig past events, ranking players, identifying future events, etc. But the primary objective in keeping the server involved in each game is financial. It is an opportunity to charge players a monthly fee for being part of the game community they have chosen.

Client/Server. Combining the ideas of Master Clock and Token Passing, Client/Server synchronization pulls more control of the distributed execution into a centralized server. Most networked computer games are implemented in this way. It provides much more accurate control of the distributed environment by releasing events to clients only when they are executable. The server can progress simulation time in accordance with a real-time clock or by some other controlled rate of advance. Most game servers progress according to a real-time clock, expecting each client to process events fast enough to remain in pace with this.

In addition to synchronizing time and events, the server can adjudicate situations in which network lag has created disjoint events like those described in the opening example of this article. The server can determine whether a weapon-hit should have happened in a perfect, non-delayed network environment. It can also make more sophisticated judgements about whether legal events should be allowed to happen based on how they will appear to the players in the virtual world. The virtual world on the server is the most accurate picture of the state of the world. Each clients has a slightly inaccurate, reflected picture of the world to work from.

Chandy/Misra/Bryant (CMB) is the common name for a popular method of imposing event synchronization in parallel and distributed DES. Each of the authors was responsible for contributing concepts to the complete implementation of the technique. Since it was developed is a DES environment, it is optimized for an event-stepped world in which events are relatively sparse within a given time period. These events are usually loaded from an initialization file or are triggered by these initial events. They are seldom received from an interactive user of the simulation.

The CMB algorithm is implemented within the infrastructure software that manages the event queues and releases events to models based on the progression of the simulation time. CMB is a key part of determining the rate at which distributed simulation time progresses. Each event posted to the simulation infrastructure by a model is sent to all of the other simulations in the parallel or distributed environment. Upon arrival, the infrastructure places each event in a local queue that is identified as belonging to the remote simulation that generated the event. Therefore, the infrastructure on each distributed machine is holding multiple queues with events segregated according to their originator. These queues can be used like a scoreboard to determine where each distributed piece is in its execution.

The simulation infrastructure orders the events in the queue from the earliest to the latest. It then locates the event with the smallest time-stamp across all of the queues. This event is safe to process because we are certain that none of the distributed simulations will generate a new event with a time-stamp smaller than this one. Executing it can not lead to a cause-and-effect error with later events in the queues.

Each simulation in the distributed environment proceeds in the same manner, generating new events, posting new events, and processing the smallest timed event in its local queues. However, it is possible for a simulation node to process all of the events from one of the queues, leaving it with no knowledge of the time at which the remote simulation associated with that queue is processing. When this happens it is not possible to select the smallest event in the queues because the next event that arrives for one of the empty queues may have a smaller time that the time of other events that are available. Proceeding with the selection of an event to process would be causally dangerous. You may arrive at a point in time ahead of the time-stamp on the next event that arrives from that simulation.

Several solutions to this problem have been considered. If we assume that the empty queue should retain some memory of the time stamp of the last event that was removed from it, this may help the situation. This could then be used when the smallest time-stamp is determined. Unfortunately, this does not help because the memory of the time on the last event processed does not move forward until a real event arrives to fill that queue. Complicating this situation, it is possible for each simulation to empty their event queues in an order such that they all become deadlocked. Imagine a 3-computer environment with the simulations labeled **A, B**, and **C**. Simulation **A** may have emptied the queue of events sent by simulation **B**. **B** may have emptied the queue from **C**. **C** may have emptied the queue from **A**. In this situation, each simulation is waiting on one of the others in an order that deadlocks the entire distributed environment.

To solve this, we need another piece of the CMB solution. Each simulation generates "null messages" or "null events". These do not carry any information about actual modeled events, they simply carry the simulation time of the next event that the simulation intends to generate. Therefore, when one of the nodes reaches a point where it is not going to generate events for some time it creates a null message that tells others where it intends to go next. Many different methods can be used to determine when a null message needs to be sent by the software. An ideal method must be independent of the specific models running on the infrastructure. It must be something that will work in all situations. The most generic and failsafe method for doing this is generating a null message after you process each real event. The null message then acts as a pre-announcement of the time that will be on the next event when it is finally generated. One drawback to this method is that it generates many more null messages than are actually necessary. Another approach would be for a simulation to generate a null event each time it sees that one of its queus is empty. This minimize null messages and insure that deadlock does not occur. But it may allow remote simulations to sit idle for some long periods of time.

There is another problematic question to be answered. How does a simulation know what the time stamp will be on the next event it will generate? If the simulation is time-stepped, the next time stamp is equal to the current time plus the size of one step increment. If the simulation is event stepped, the next event could be at any time in the future. Luckily, even event-stepped simulations are written with some understanding of their inherent step-size. It is possible for an event stepped simulation to identify the soonest it is possible for the software to generate the next event. This is usually based on specific details about how the simulation is structured. Therefore, after generating all events for time t, the next event must be generated at time $t+1$ (or more strictly $t+dt$). The time step size or simulation clock counting increment dt is defined to be the "lookahead" of the simulation. The lookahead value is then added to the current time to determine the time stamp that will be delivered in the null message.

As the result of experimentation it was apparent that the rate of progress of distributed virtual worlds was directly effected by the size of the lookahead of the simulations involved. Larger values allowed the federation of simulations to move ahead more efficiently than smaller values. But large values also reduce the flexibility of the simulation algorithms. Therefore, the selection of lookahead is always a trade-off decision between modeling flexibility and processing efficiency.

Example Algorithm: Chandy/Misra/Bryant Conservative Synchronization

```
while (!done) {
        waitUntil(each FIFO queue contains at least one event);
        getEvent(smallest time stamped message in all queues);
        simTime = eventTimeStamp;

        <Process this event>

        sendNullMessage(time stamp = simTime + lookahead);
}
```

Aggregate Level Simulation Protocol. In the early 1990's the military was searching for an alternative method for synchronizing distributed wargames. They had been using the Master Clock method and were suffering from some of its limitations. The CMB synchronization algorithm was considered and experiments conducted. The time-stepped nature of the wargames indicated that some very serious performance improvements could be made for the specific Joint Training Confederation (JTC) that was being created.

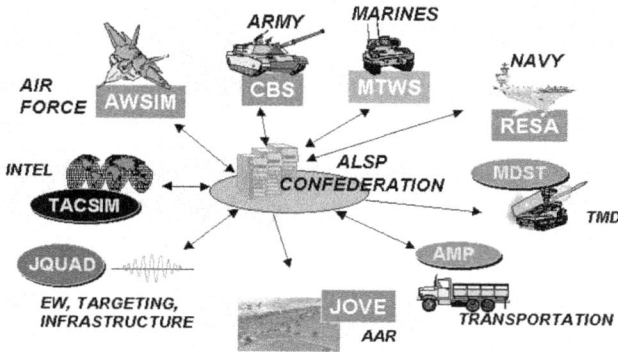

Linking Military Wargames.
The Joint Training Confederation

Since each simulation was time-stepped, the lookahead value was very obvious. However, each simulation was responsible for many hundreds or thousands of objects and each of those executed and generated events at each time-step. Therefore, a pure CMB algorithm would generate a null message after each real event, essentially doubling the number of event messages being sent through the network. Since several hundred units all executed one or more events at the same time stamp, the information carried in the null message was very redundant. Therefore, a modification was devised that eliminated thousands of null messages, but retained the power of the CMB algorithm.

The infrastructure was modified so that it no longer considered the time-stamp on each event in its search for the lowest distributed simulation time. Instead it would consult only a new, special type of event—the "Advance Request" event. Before stepping to the next time-step, each wargame would post an "Advance Request" to the infrastructure. These would be exchanged between all members of the virtual world and evaluated to determine the lowest time that was being requested from all of the simulations. Each instance of the infrastructure would then turn around and give its simulation an "Advance Grant" for the smallest time. Simulations that wanted to advance to that time would do so. Others would wait for an advance grant for the time they had requested. Those that did have events for the time granted would process them and post another "Advance Request" for a time further in the future than the last one. As this continued simulation time would advance, each simulation would be able to process its events, and each simulation would have control over the rate at which the distributed time progressed.

This approach to event management and synchronization became part of the Aggregate Level Simulation Protocol (ALSP) that is used to tie distributed wargames together. More information on this protocol and project is available at http://ms.ie.org/alsp/.

OPTIMISTIC SYNCHRONIZATION

In an effort to squeeze even more speed and efficiency out of distributed simulations, a group of scientists began experimenting with ideas in which the simulations are free to process events as fast as possible. But they are still required to retain the cause-and-effect relationships between events and objects spread across the distributed simulation processes. This was pioneered by David Jefferson at the University of Southern California. It was clear that every implementation of conservative synchronization required faster simulation nodes to remain idle while waiting for one of the slower members to process events. This was wasting CPU cycles that might be used to process events locally. In fact, processing these event "ahead of time" might allow the entire simulation federation to finish its job much quicker. Unfortunately, processing events ahead of time may break the cause-and-effect relationship between events and objects on different computers. Some method was needed to use the available CPU cycles, progress simulation time efficiently, but retain cause-and-effect relationships. Thus was born Optimistic Synchronization.

We should make it clear that these ideas were being formulated for DES used to analyze data and process events that were pre-defined in the computer model. Interactive users were not involved during the execution of those simulations. We will add interactive users to the mix after we have described how Optimistic Synchronization works for non-interactive simulations.

Time Warp. David Jefferson created an approach called Time Warp that implemented his ideas for optimistic synchronization and which has become the foundation for most of the work in this area. Under Time Warp, each simulation on the network operates as if it is running entirely by itself. Every event that arrives on the computer, whether from a local data file or a network message is delivered immediately to the simulation models for execution. The models are assuming that both the local and distributed events will arrive in the correct time-stamp order. If this assumption were 100% correct, then the cause-and-effect relationship would not be violated and maximum use would be made of the CPU cycles available. However, this is assumption is never 100% correct. In many cases the events do arrive in the wrong order and events are processed incorrectly. When this occurs, some mechanism must be used to redo the calculations in the correct order. The "undo" and "redo" operations required to accomplish this will require CPU cycles. Therefore, Time Warp is a balancing act between the CPU cycles captured for processing events early and the CPU cycles lost when fixing problems caused by this approach.

We will explain the operations of the Time Warp mechanism with a combat example. Imagine a federation of three simulations. One handles all army units and we will call it Ground. Ground is responsible for tanks, soldiers, trucks, and surface-to-air missile batteries. The second handles all airborne units (Air)—fighters, bombers, tankers, electronic warfare, and AWACS. The third handles all sea-going units (Sea)—carriers, battleships, submarines, and supply ships. In a particular scenario these three simulations are using the same computer hardware, but each has a very different software model and must manage widely varying numbers of units. Ground must simulate the operations of 10,000 units, Sea must simulate 1,000 ships, and Air must simulate 100 aircraft. In this situation it is almost certain that the three CPU's will have very different loads imposed on them and will progress forward at different rates.

The Air simulation launches a flight of aircraft to bomb a carrier at sea. That simulation flies the planes to the projected (via dead reckoning) position of the aircraft carrier and drops the bombs on that location. However the Sea simulation has not processed to that point in time yet. So it accepts the posted bombing event and places it on the queue for future processing. The Air simulation continues to process, flying the aircraft back to the base and landing them. At some point, Sea arrives at the simulation time of the bomb release, recognizes that the bombs were dropped on the correctly predicted location of the carrier and processes the explosion and resulting damage. This is a best case scenario for the Time Warp mechanism.

However, assume that the Ground simulation has been processing through time very slowly. Long after the bomb has been dropped, the Ground simulation processes the flight event that took the aircraft past a surface-to-air missile (SAM) battery on its way to the target. The SAM model recognizes that it can detect the aircraft, shoots a missile at it, and hits it. This all occurs at a simulation time prior to that at which the bomb was dropped. When the shoot-down event is posted to the network and arrives at the Air simulation, the Air simulation must "undo" all of the events it executed after the shoot-down time and recalculate the sequence with the shoot-down event inserted at the appropriate time (Figure 6.1).

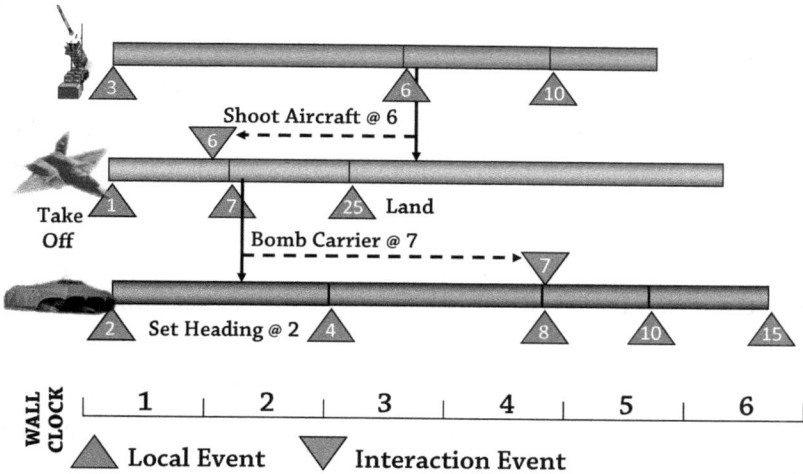

Figure 6.1. Rollback in a Time Warp Simulation

To support this, Time Warp and Optimistic Synchronization methods must record all changes to state variables. These records are then reapplied to the aircraft object in reverse order until it is "rolled back" to a state prior to the shoot-down event. The Air simulation then inserts the shoot-down event in the event queue and reprocesses the sequence. This time through the aircraft discovers that rather than having a successful mission that can be told around the officer's club, it is shot down, the pilot ejects, and must be rescued before he is lost at sea. Since the aircraft was rolled back and shot down, there must also be a mechanism for retracting events that were generated in the first pass, and must be undone in the second pass. Events that were generated by the aircraft are stored, just as state changes were, so they can be retracted (rolled back) as well.

Object state rollback is relatively easy because it is entirely contained on the local computer. But retracting events posted on the network is accomplished by posting "anti-messages" or "anti-events". These travel to the remote nodes and are used within each instance of the infrastructure to locate the original event and remove it from the queue. If the original event has already been processed, the anti-message will cause that simulation to rollback as well. It must rollback and reprocess its sequence of events with the retracted event removed from the queue. When the rollback of one simulation causes another to rollback as well, it is referred to as "cascading rollback". As the name implies, in some situations the cascading effect can go much further than just one or two objects. It may trickle all through the entire distributed simulation.

As chaotic as this sounds in human terms, within the computer it is merely an annoyance that costs CPU cycles and prevents the parallel or distributed simulation from finishing as soon as was hoped. The processing of events, anti-events, state changes, and rollbacks are just computer operations designed to study a problem and arrive at an answer as soon as possible. There is no need for an operator to see this non-monotonic progression of time. There is one major fly in the ointment—the state and event data that is stored in the computer in preparation for rollback consumes valuable RAM. Some mechanism is necessary to release this information when it is no longer needed and reclaim the memory it is using. To do this we must be able to select some simulation time beyond which roll back can not occur. This is where the CMB and ALSP concepts of a unified federation time are valuable.

If we evaluate the current processing time of all of the distributed applications, one of them has the smallest simulation time-stamp. That time represents a point at which no additional events can be generated with a smaller stamp. Therefore, rollbacks can not push any member of the federation back further than that time. Any saved state and event information prior to that time can be safely reclaimed and used for other operations (perhaps to store the newest state saves that are being generated).

Unfortunately, because this is a distributed simulation there may be many events that are in-transit between the sender and the receiver. This requires the use of some rather intricate algorithms to identify the true smallest simulations time in the distributed world. Descriptions of several of these algorithms are beyond the scope of this article and can be found in Fujimoto's book. This lower time boundary is known as the Global Virtual Time (GVT). GVT can be viewed as the unified time of the distributed simulation (Figure 6.2).

There are additional details on the implementation of Time Warp and Optimistic Synchronization that are beyond the scope of this article. Interested readers should consult *Parallel and Distributed Simulation Systems* by Richard Fujimoto. It is the definitive source of information on this topic.

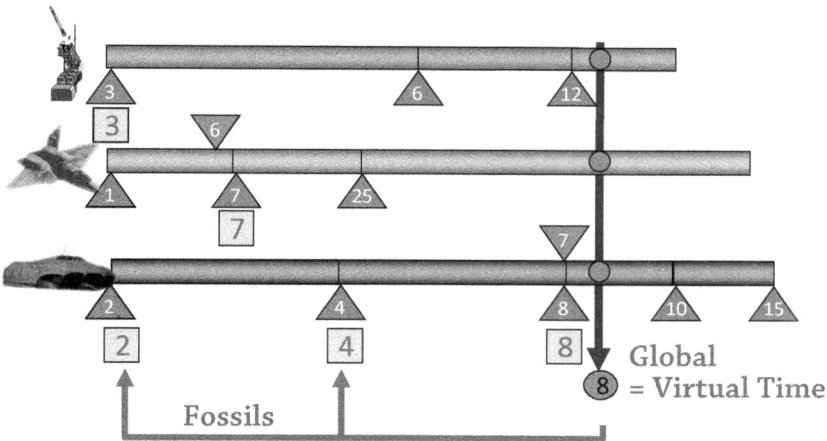

Figure 6.2. Identifying Global Virtual Time

Man-in-the-Loop Time Warp. Early in the discussion of Time Warp we pointed out that it had been developed for analytical applications that did not include interactive man-in-the-loop operators. Some work has been done to apply this technique to interactive simulations and gaming environments.

But, in general, the insertion of an interactive user negates most of the advantages of this approach. Human players expect the simulation to progress at real-time, so the advantage of processing events as fast possible is reduced to remaining synchronized with real time. But since the simulation is not allowed to fall behind real time at any point, the computer hardware must be capable of processing all events and rollbacks at a real time rate. This may require more powerful hardware than is necessary for conservative synchronization.

Interactive players also can not witness the pre-processing and rollback of events in the simulation. The simulated world must be revealed to them only as events are guaranteed not to be rolled back. Therefore, all interactive users must "ride GVT". Meaning that they see only events and states as GVT advances over them. Since the player may insert new events at any time, he becomes the default slowest simulation in the federation. The progress of GVT is totally governed by the progress of the players, which is defined to be at real time.

The one remaining advantage of implementing Time Warp in interactive simulations is that it allows processors to work on large clusters of events before the player sees them. This has the potential of avoiding simulation slowdown at critical times when lots of events occur simultaneously. This work may be entirely pre-processed by the Time Warp simulations. However, it is probable that during large clusters of events the interactive player will want to insert new events reacting to what he sees happening. This will cause rollback at the worst possible time.

In general, Optimistic Synchronization is not the best solution for interactive simulations used for military training or computer gaming.

RECOMMENDED APPLICATIONS

After describing many of the synchronization methods used for distributed virtual worlds, it is appropriate that we identify which are best for different types of applications. This list can not be all encompassing, but it can categorize many of the major types of simulations.

Synchronization Method	Application
Independent	Flight Simulators
Token Passing	Card Games Turn-Based Strategy Games
Master Clock	Model-to-GUI Coordination
Client/Server	3D Shooter Games Real-Time Strategy Games
CMB/ALSP	Distributed Wargames
Optimistic	Large Analytical Simulations
Regulated Optimistic	Analytical Simulation with High Interactions

Chapter 7

N-DIMENSIONS OF INTEROPERABILITY

A round 411 BC, Plato wrote "The Republic" and in one dialogue challenged his Greek pupils to imagine a place where people lived in a cave and were chained to the floor such that they could see only the wall of the cave and not the opening. Their entire perception of the world were the shadows of objects and activities that took place outside and were cast on the wall of the cave in front of them. In this world the shadows would be, for the chained audience, their perception of the real world. They would believe that there was no more to the world than what they were seeing. He illustrates the point that reality is defined by what we are able to perceive and our world expands as our perception expands.

In 1884, Edwin Abbott wrote a small book *Flatland: A Romance in Many Dimensions*. He portrayed a world of only two dimensions and described the creatures living there, their environment, and the methods of conducting their lives. These creatures were entirely ignorant of the third dimension and interactions with a three dimensional creature appeared magical to them.

In both cases, the characters in the stories are robbed of the perception of one of our fundamental dimensions of reality. These illustrate how a world can be real with fewer dimensions than we deem natural—the simulation community is in a similar situation. We have spent several decades chained in a cave of few dimensions and now find ourselves set free to explore the many dimensions of the shared synthetic world. This paper attempts to illustrate what some of these dimensions are and how they are related.

SIMULATION EVOLUTION

Though simulations have been used by the military for analysis and training for centuries, it was not until the advent of computer networks that the issue of integrating multiple heterogeneous systems emerged. Prior to that, it was naturally assumed that each simulation was a virtual world unto itself, and that information exchange was a laborious, human-driven process. Analytical simulations were the first to crack the "interoperability barrier" by creating simulations that were specifically designed to accept the output of another model as the input for beginning a different stage of analysis.

The virtual community created the Simulator Networking (SIMNET) systems with their associated protocols for joining multiple copies of the same simulator into a common synthetic world. Those these were intended for homogeneous simulations at a single location, they pointed the way for more ambitious projects. Distributed Interactive Simulation (DIS) protocols followed, adding the ability to join a large variety of simulations and to do so across wide area networks.

As this was developing the constructive community was creating the Generic Data System (GDS) to bridge the gaps between constructive simulations.

These allowed widely different simulations to exchange world data and deposit it in a common data repository and analysis system. This was followed by the Aggregate Level Simulation Protocol (ALSP) which created a more dynamic environment and provided distributed management functions.

The High Level Architecture (HLA) promises to join both of these communities and allow them to work together at a level not attainable before. Simulations from many fields: staff training, crew training, live training, analysis, engineering, and testing will be able to take advantage of a single architecture for distributed simulation.

The next step in this evolution is the integration of real world C4I systems into the synthetic world. Projects like the Modular Reconfigurable C4I Interface (MRCI), Tactical Simulation System (TACSIM), and other experiments are illustrating the fact that these command systems can be extensions of the distributed simulation paradigm. TACSIM supports dedicated interfaces between simulations and C4I systems and MRCI promises to extend this to the same generic level available to simulations through the HLA Run Time Infrastructure (RTI). With this continuous expansion of the dimensions of interoperability it is probable that the integration of other forms of combat computer systems will follow: radar trackers, radios, communication stations, operational sensor systems, and even combat aircraft. As real systems, particularly C4I systems, settle into the use of standard message formats and data exchange methods, the level of interoperability between these different fields will continue to increase.

The era of dedicated, stand alone simulation models is coming to a close. The class of problems that can be addressed with today's technology far exceeds the ability of a single model to capitalize upon. Distributed, cooperative simulation solutions that can still operate independently are going to be the systems of the foreseeable future.

A topological illustration of each of the levels of interoperability is shown in Figure 7.1.

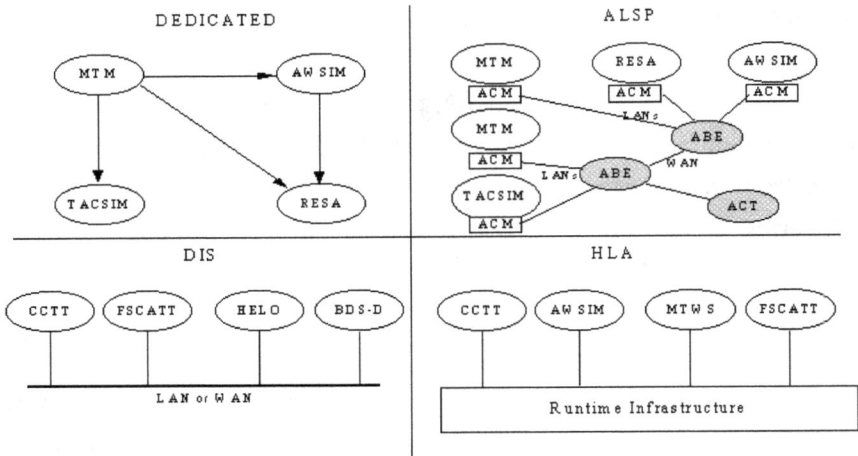

Figure 7.1. Simulation Interoperability Techniques

INTEROPERABILITY DEFINITIONS

As interoperability evolves we find ourselves with several definitions of what it is. These vary according to their intended audience and the point-of-view of the author, but three prominent and compatible definitions are:

"Two or more simulations/simulators are interoperable for a given exercise when their performance characteristics support a fair fight to the fidelity required by the exercise."—**COL James Shifflet, US Army STRICOM**

"The ability of a [...] simulation to provide services to, and accept services from, other [...] simulations, and to use the services so exchanged to enable them to operate effectively together."—**HLA Architecture Management Group**

"A domain-specific software architecture is, in effect, a multiple-point solution to a set of application-specific requirements (which define a problem domain)."—Will Tracz, **Domain Specific Software Architecture**

These represent the operational, network, and architecture points-of-view of the same problem. Each sees different aspects of the problem as primary, and each applies a unique solution to the problem. But they are all attempting to achieve similar results.

DIMENSIONS OF INTEROPERABILITY

We will attempt to enumerate and describe the various dimensions of interoperability that exist or are appearing on the horizon in requirements for new simulation systems. It is clear that the emergence of new technologies results in the creation of new dimensions in interoperability. Each capability brings with it the possibility of extending the existing simulation in new directions, so we do not mean to imply that this list of dimensions is comprehensive. Some of the dimensions are often implemented together in the application realm so that they appear as two or more unique dimensions only during analysis. The twelve dimensions described below may fold together to form fewer dimensions with richer characteristics when actually implemented in a simulation, but each of them does have unique properties which justify its independent identity here.

Expanding Heterogeneity

When a simulation is built to solve a single, self-contained, bounded problem it is difficult to conceive of dimensions of interoperability as a significant problem. But, as simulations are connected to others of very diverse natures and the whole applied to different problems in each domain, the interoperability problems multiply. Current desires to join all forms of simulations and military computers create the high dimension environment we are working to capture and describe.

Military Dimensions. Within the military domain, the first class of dimensions are naturally those of Military activities (Figure 7.2). The services have spent the last two centuries defining and separating themselves from each other, creating their own unique identity. Simulation interoperability is now asking them to identify where they are the same so that they can operate together. We must deal with the differences that have been erected over two centuries between Military Organizations: Army, Air Force, Navy, Marine Corps, Coast Guard, Special Forces, Intelligence Organizations, the Department of Justice, and the Drug Enforcement Agency.

MILITARY

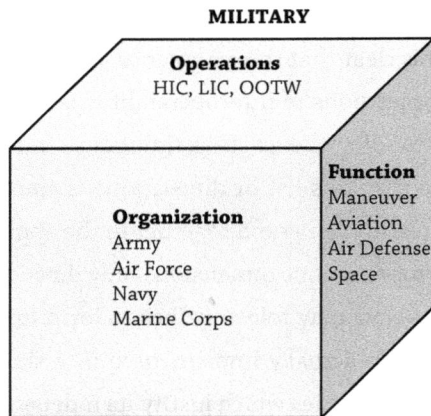

Figure 7.2. Military Class Cube

Strictly speaking each of these is not a part of the DOD. However, in the modern world they are called upon to interact and support one another to implement the domestic and foreign policies of the nation to which they belong. So it may be just as important for the DEA to exchange information with the Defense Intelligence Agency as it is for the Army or Navy.

The second dimension in the military class is that of Function. Within each service, specific functions have been isolated and have created their own operational environment and culture. They have specialized themselves in order to better perform their mission, but this makes them less like the other functional areas and generates barriers to communication and interoperation which must be surmounted. Some of these functional areas include: Maneuver, Intelligence, Air Defense, Artillery, Engineering, Aviation, and Combat Service Support.

The third dimension in the military class is Operations. Simulations exist which correctly describe the interactions of forces in large scale combat, but incorrectly model small contingency operations. In these cases it is necessary to join two simulations which portray specific operations accurately. Some of the operations that must be interoperable are: High Intensity Conflict, Low Intensity Conflict, Operations Other Than War, Law Enforcement, Intelligence Collection, Diplomatic Negotiations, and Psychological Operations.

In any given simulation there may be characteristics from all of these dimensions modeled, though it appears to be common to find certain characteristics modeled together, such as Army-Artillery-HIC or Special Forces-Aviation-OOTW.

Interaction Dimensions. The next class of dimensions is Interactions with simulations (Figure 7.3). These are labeled as the Level, Method, and Extent of interaction between the simulation system (software and hardware) and the humans who are using it.

INTERACTION

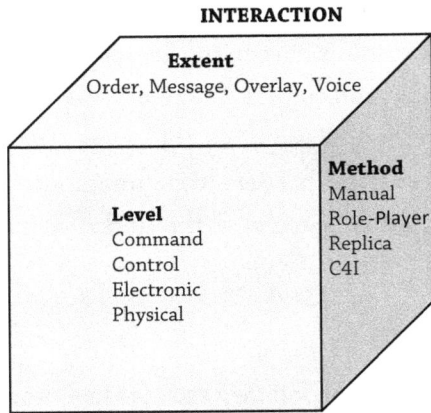

Figure 7.3. Interaction Class Cube

The Level of Interaction dimension describes the training audiences mental perspective when working with the simulation. At the Command level the trainees are interested in managing large groups of entities that may be aggregated into units. He/she is interested in telling the units what to do and seeing how they responds to those commands. This level is most common in constructive, staff training simulations. For finer detail there is the Control level in which the trainee is actually controlling a simulated device and the intent is to wrap the simulation around them without their noticing that it is not real. This level is most common in virtual vehicle simulators. In the Electronic level the trainee has an electronic connection to the simulation. A computer, radio, radar screen, or other device is their window into the combat world and their physical surroundings are primarily insignificant to their mission. Lower still is the Physical level in which the trainee is actually in a combat environment which is real up to the point of lethal exchanges. A soldier at the National Training Center is in this position, the surrounding environment has everything to do with his mission and human eyes and ears are the window into the world. Finally, there is a Listen-Only level in which the soldier is expected to receive information to analyze and disseminate. He/she is not expected to react to the incoming events, merely to record them.

The Method of Interaction is the second dimension in this class. It describes the actual tools that are used to connect the trainee to the simulation. The Level dimension above often implies the use of certain tools based on past experience, but the focus there is on the humans mental perspective. Here we are interested in connecting the tools so they can be used in the simulation. The Manual method is the most ancient. Here the soldier is directly manipulating the battlefield as it lays in front of him as with board wargames, chess, and computerized versions of these games. The Role-Player method is used to isolate the trainee from the artificialities of the simulation, placing a human between the trainee and the simulation system. This person plays the part of an actor maintaining the realistic environment for the trainee. The Replica method constructs a shell around the trainee so that the simulator looks and feels like real combat equipment. This is the method of virtual vehicle simulators. The C4I System method is a hybrid of the Replica and Role-Player. It attempts to provide the same realistic interaction to command and staff players that has been experienced by vehicle crews. However, it must interact with the trainee at a much higher mental level, appearing to converse with him/her just as the role-player did in the past.

The Extent of Interaction is the dimension that describes the detail to which the above methods are carried. This dimension determines the depth, richness, and realism that the trainee experiences in the Method dimension. The poorest Extent is to provide order parsing where the simulation understands only the syntax developed specifically to operate it. This tends to corrupt realism and call for the use of a role-player to negate the effects. The Message Parsing extent allows the simulation to receive stimuli from the trainee in a form that comes from the real world. For staff trainers this may be tactical messages. For crew trainers this may be steering and firing signals from real equipment. Overlay Parsing provides an extent which allows the exchange of much more complex information. It augments the information content to include pictorial information as well as text. Voice Recognition allows both staff and crew trainers to control the simulation with the same stimuli that they would use to control real world assets and equipment. Adding Voice Generation closes the loop and allows the simulation to respond to human voice input in kind.

System Dimensions. The System class of dimensions deals with the implementation of the simulation (Figure 7.4). Different capabilities in interoperability are dependent upon how a simulation is actually created and how the real world is represented. The first of these dimensions is the classic Level of Representation. This has been divided conveniently into the categories: Constructive, Virtual, Live, Analytic, Testing, and Engineering. These usually describe the structure of the model as chosen to provide the answers or functionality needed by its creators and customers. Simulations at different levels of representation face the problem that information available on events or objects is not compatible with other simulations. In the best case, the information in one is an aggregate or disaggregate of the information in another, and there is some hope of joining them. In other cases the information is in no way related, functionality provided in one was considered inappropriate or insignificant in the other. In some cases, this also results in a situation in which there is no mission need to join these systems.

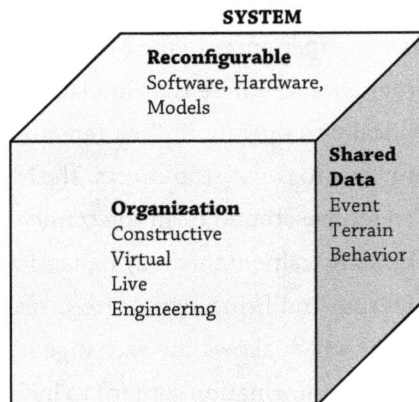

Figure 7.4. System Class Cube

The next dimension of this class is Shared Data. In the past, the information exchanged between simulations has been related to specific events. This has left environmental and behavioral data uncorrelated. For example, the terrain in one simulator may look very different than the terrain in another, which may be due to differences in the data sources or differences in the interpretation of data sources. The same is true of the visual models of objects, light intensity, and shadowing in different simulations. Significant differences may also exist in the implementations of behavior, making it more effective to attack a unit in simulator A rather than in B because it is known that simulator A has a more passive behavior, or integrates the effects of morale and fatigue, where simulator B does not.

The last dimension in this class is Reconfigurability. This defines the ability of the simulator to be used in more than one configuration. At the hardware level this speaks to the ability to host a whole family of simulations on different sets of computer and communication networks. This may allow the instantiation of a capability at one location, or a rotation of instantiations to support varying training missions. At the software level this describes the ability to change the simulator to represent more than one vehicle, level of representation, method of interaction, etc. Most simulators are built to operate in a single configuration with one customer and one basic mission. It may be possible to extend this to allow a single system to serve multiple missions and customers.

Programmatic Dimensions. The Programmatic class discusses issues external to the actual use of the simulation and focuses more on its design and architecture (Figure 7.5). The first of these is the Domain Engineering dimension. The architecture of a simulation can be constructed uniquely to serve a single customer, or it may be part of a larger interoperable architecture that joins the proponents of many simulators. If this dimension is exploited, a large of set of simulation interactions are satisfied in the design of the system and do not have to be addressed separately, and less effectively, during individual system implementation.

PROGRAMMATIC

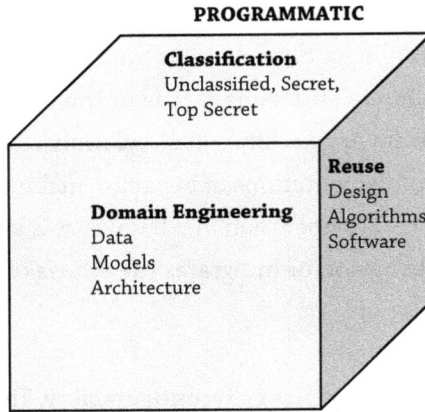

Figure 7.5. Programmatic Class Cube

Reuse is the second dimension of this class. It measures the degree to which simulation capability is shared among multiple simulations. Where reuse is high there is the potential for greater consistency in both the environment and behavior of simulators as described in the Shared Data dimension above. Reuse may be at several levels: design that defines capability, algorithms that control activities, software that implements algorithms, and hardware that hosts software.

Finally, Classification divides simulations according to the sensitive nature of the information and capability they portray. This is related to the Military Function and Shared Data dimensions described above, however, due to its unique position it must be addressed and bridged separately. In the past, it has been rare for simulations at different levels of classification to work together, though there are instance of this. Where it is allowed, the classification dimension has always been significant and separate from other dimensions of interoperability.

Graphic Representation

The sections above have organized the dimensions into classes to aid in describing them. Here we would like to give a graphic representation to these dimensions to aid in their manipulation, addition, and organization. We are providing two different pictures of the interoperability space, the first is a traditional class diagram (Figure 7.6) which implies that each dimension shares some base characteristics with the others in the class. This may be true, but it is too early to determine whether the correct class groupings have been reached. It also begs the question of what these shared characteristics are, a question we are not prepared to answer in this paper.

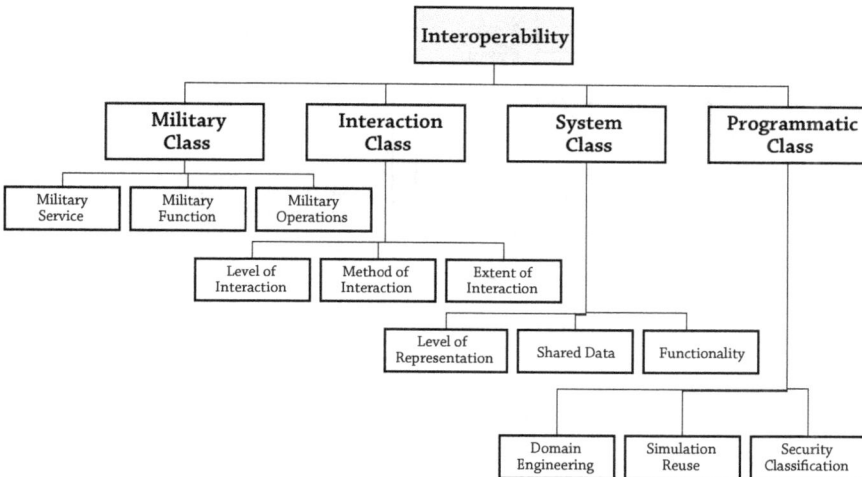

Figure 7.6. Dimensional Class Structure

The second diagram (Figure 7.7) is derived from a four dimensional hypercube that has been unfolded into three dimensional space. This was first described in 1940 by Robert Heinlein in his classic short story "And He Built a Crooked House". It implies that there are definite adjacencies between dimensions as shown in the three dimensional picture and there are adjacencies that occur when all dimensions are integrated into a single system, similar to the effect of folding the 3-D diagram into a single 4-D cube. This diagram may be more useful for illustration of relationships than for general work on improving relationship definitions.

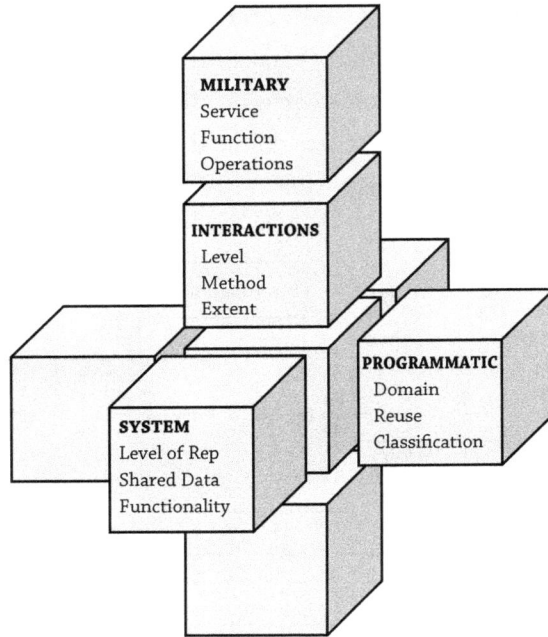

Figure 7.7. Dimensional Hypercube

Dimensional Algebra

In his 1976 book *Theory of Modeling and Simulation*, Bernard Zeigler developed a formal specification for simulations. It is desirable to have a similar specification for the dimensions of interoperability to aid in the definition of ideas and their manipulation. This has proven to be quite difficult and the specification provided here is the shadow of a beginning.

Interoperability, I, is a function of many different variables. In this paper these are grouped and characterized as dimensions, d. The dimensions are grouped into classes, c, to illustrate relationships and potential dependencies. Each dimension has discrete instances, i, and each instance consists of a set of variables, v, as illustrated in Equation Box 1.

A specific instance of interoperability, such as marine corps models, would be written as:

I(c-military(d-service(i-marines()))).

A specific variable within the marine corps model would add one more level of detail than has not been discussed in this paper:

I(c-military(d-service(i-marines(v-org structure)))).

Equation 1. Dimensional Algebra Illustration

I = f(dimensions) =>
  I(c-military, c-interaction,c-system,c-program) =>
    c-military(d-service, d-function, d-operations), c-interaction(d-level, d-method,
    d-extent), c-system(d-rep, d-share, d-reconfig), c-program(d-domain, d-reuse,
    d-classification) =>
      d-service(i-army, i-af, i-navy, i-marines, ...), d-function(i-mvr, i-css, i-intel, ...),
      d-operations(i-hic, i-lic, i-ootw, ...), ...=>
        i-marines(v-org structure, v-culture, v-mission, ...)

I = Interoperability
c-* = four classes of interoperability described in paper
d-* = individual domains within each class
i-* = discrete instances within each dimension
v-* = simulation variables which define instances

STRUCTURED APPROACH

The fields of biology and chemistry struggled to impose structure on their problem space for centuries. Some of those taxonomies are useful to today's simulation community as we try to do the same thing. As we seek to define object models of the simulation, federation, and domain we should recognize that the biologic community has developed a hierarchical taxonomy to group animals and plants so they can work with them in an organized fashion, and the chemistry community has created the table of atomic elements, both of which perform a similar function to our object models.

The simulation community has sought to define a structure that is useful for creating models of different problem spaces, as in Zeigler's book. One of the difficulties of creating a taxonomy of simulation is caused by the breadth of the field. We are apt to create a simulation of any process in the world, a problem space that would require a World Object Model (WOM) to completely support it. Even within the military training community there is a vast ocean of possible modeling needs. The use of a domain object model and domain architecture is a step that is being taken within the Joint Simulation System (JSIMS) program. This will attempt to pull together the divergent, service-specific models and give them a level of commonality in the architecture which will support interoperability and reuse in other dimensions.

CONCLUSION

The dimensions of simulation interoperability will continue to increase as the demands we place on them increase and the military continues to computerize its forces. The need to define, structure, and manage these dimensions is becoming more evident all the time. Only by organizing this aspect of simulation can we hope to control the development cost, harness the potential of software reuse, and implement domain architectures.

The dimensional algebra described is only the seed of the type of formal definition that is needed to aid in managing and manipulating dimensions of interoperability. When this is coupled with a structured object modeling approach it will allow us to identify simulation capabilities that are useful beyond the military training domain (Table 7.1). Though often referred to as technology transfer, these civilian similarities are an acknowledgment that the US military is an integrated part of the American society and is attempting to solve some of the same problems faced in many other areas.

Table 7.1. Civilian Applications

Military Function	Civilian Application
Combat Service Support	Product Distribution, Medical Services, Social Services
Command, Control, and Communications	Telephone System, Internet, Electronic Banking
Maneuver	Emergency Management, Prison Riots, Highway System
Intelligence	Business Strategy, Market Analysis
Air Defense	Federal Aviation Administration
Engineers	Construction, Emergency Management, Civil Services
Electronic Warfare	Radio and Television Spectrum Analysis
Aviation	Airline Route Planning
Space Operations	Communications Satellites

Chapter 8

COUNTER TERRORISM SIMULATION

T he terrorist attacks on the World Trade Center demonstrated that we are not completely safe from foreign attacks on our home soil. For decades, the United States has focused its military and intelligence capabilities on enemies beyond its own borders. But, it has become incredibly clear that our enemies have the ability and the determination to reach through our defenses and strike at critical assets here at home. Our political, military, and intelligence resources are now defending our interests by defending our own homeland.

Homeland security has already taken on a new importance in government. Prior to September 11th most people on American soil felt safe from foreign attack. Today, we look to our government for protection from attacks like those on the World Trade Center and Pentagon. Our government in turn looks to its servicemen and contractors to create and operate those defenses.

Modeling and simulation can be part of this new mission. It can provide knowledge, understanding, and preparation against future attacks. From 1945 until 1995, the M&S community has devised algorithms and systems to help us understand the threat from large-scale conventional, nuclear, chemical, and biological attack. These algorithms and systems were not laying on the shelf waiting to be put into service, but rather, had to be created piece by piece over a period of many years. We now face the need for a new set of tools and must begin creating those tools now if we are to be prepared for future terrorist threats. Within the M&S community our responsibility is to create algorithms and systems that can contribute to this fight, because terrorism will be to the 21st century what the Cold War was to the 20th.

THE EVOLUTION OF NEW SIMULATION MODELS AND TOOLS

The simulations necessary to address the complexity of the terrorist threat must evolve just as the current simulations of traditional combat have evolved over decades. A few organizations have been heavily involved in quantifying the terrorist threat and identifying the relationships that hold it together. But, the largest part of the simulation community has been focusing on Desert Storm-like scenarios. Therefore, most of the military and intelligence modeling profession faces a steep learning curve in shifting to this new mission.

Wargames. Creating manual wargames allows us to identify objects, attributes, events, and relationships that are relevant to the problem we are trying to solve. Working with maps tables, documents, and counters allows modelers to focus on the structure of the problem rather than the structure of a computer language of software tool.

System Dynamics. System dynamics models are designed to capture nuggets of knowledge and hold them together so the complexities of their relationships can emerge. Since Jay Forrester pioneered this concept in the 1950's, models have been created to help us understand the complexities inherent in all types of dynamic systems. The key here is gaining an understanding of how terrorist organizations work.

Operations Research. Once we understand how a dynamic organization works, OR can be used to identify the optimum methods of countering it. We can use these models to identify its weakest points and to quantify our responses to its actions.

Training Simulations. Understanding how the target operates and how best to respond to it, we will then be in a position to capture this in interactive training simulations that can be used to teach people to take action against the new terrorist threat.

Each of the model classes above contribute essential information to our understanding of a problem and to our ability to accurately capture it in a simulation.

TERRORIST ORGANIZATIONS AND ACTIONS

Like a foreign nation, a terrorist organization is a complex system composed of many different interlocking components. Each component and each relationship between components is a potential target for our countermeasures. A key part of creating M&S tools to address terrorism will be determining which actions work best against which targets.

A simplified view of a terrorist organization is given in Figure 8.1. The **Command Nucleus** holds the organization together and is the ultimate source of direction and orders. However, it is not the sole conduit for all information, money, and weapons. The **Field Cells** operate under the direction of the Command Nucleus, but also possess a good deal of autonomy in accessing financial assets and weapons. **Financial Assets** are provided by **Sympathizers** and by the Command Nucleus. These allow the field cells to function on a day-to-day basis and give them the resources necessary to carry out a mission. Weapons are purchased openly or on the illegal market. **Host Nations** are those political and geographic powers that protect and host the terrorist organization. This support is particularly important when an organization is small and just getting started. It is also essential for providing a location for recruiting and training operatives. **Communications** are an essential part of keeping the organization together and coordinating its actions.

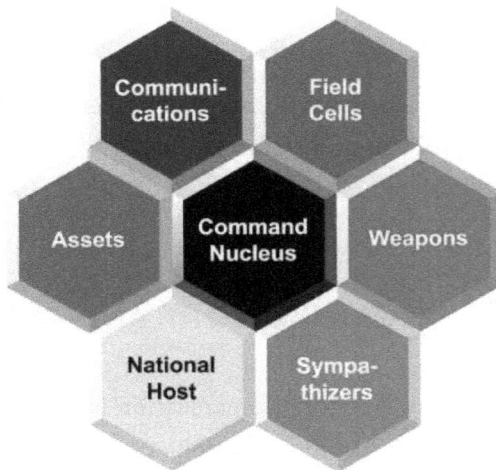

Figure 8.1. Simplified Terrorist Organization

Each piece of the organization presents a unique target for our countermeasures. Each piece also exists in multiple domains (Figure 8.2). The Com-

mand Nucleus and Field Cells are physical assets that can be reached with traditional military and police forces. Communications exist in an information
domain that is accessible to our intelligence resources. A Host Nation exists
in the political domain that must be addressed by political forces. Sympathizers exist in the physical domain, but the real target of interest here is one of
culture and community. Financial assets exist in a unique domain accessible
only to financial organizations. Finally, captured people, organizations, and finances are delivered to the legal system to be eliminated.

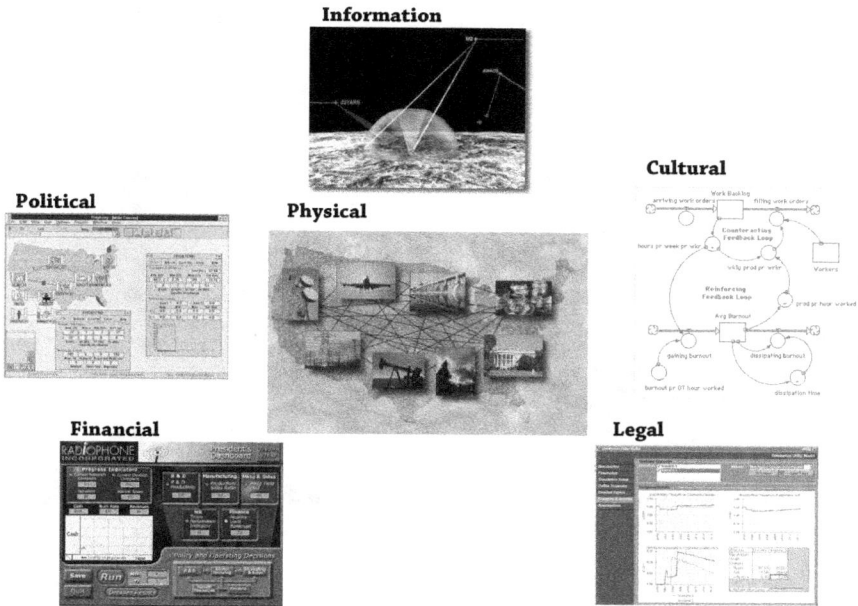

Figure 8.2. Domains of Terrorism

This diversity calls for the employment of a wide variety of assets. It is
not immediately clear what all of the relationships are between these pieces
or how the application of one countermeasure will impact the entire organization. There remains a great deal of research and experimentation to unveil

these relationships. In the following section we will discuss some of the countermeasures that can be taken in each domain and the impacts this will cause in other domains.

SIMULATION'S APPLICATION TO COUNTER-TERRORISM

The different domains that must be explored through the tools and methods described above are shown in Figure 8.3 and are described in the section below.

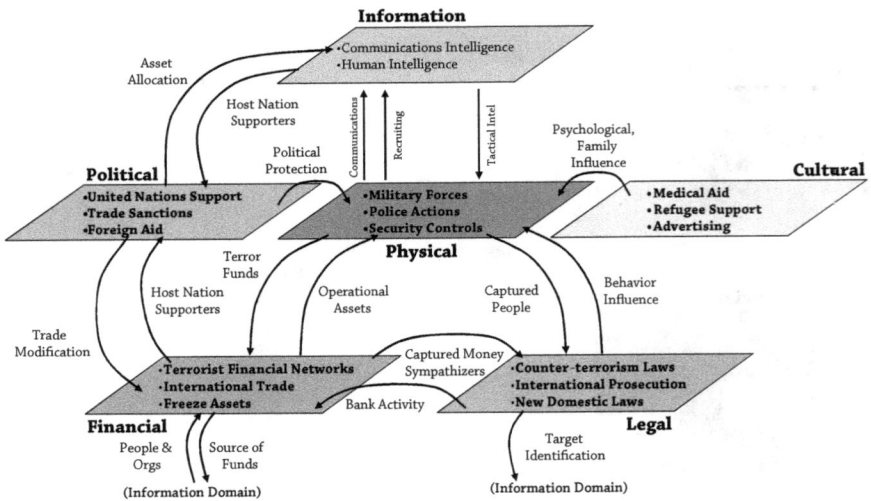

Figure 8.3. Cross-Domain Interactions

Military Simulation

The military portion of the scenario includes actions like those conducted in Afghanistan in search of Al Qaeda. A traditional combat simulation in which aircraft bomb targets, helicopters deliver special forces, infantry engage in fire-fights, and surveillance aircraft search for targets can be used for the military portion of the mission. Existing wargames and semi-automated forces systems can be adopted to meet this need.

But such a system does not represent the intelligence assets and activities that have gathered clues to the identity, location, activities, and plans of the targets. This must be carried out in a different federate.

Intelligence Simulation

Intelligence simulations represent all of the intelligence, surveillance, and reconnaissance assets available to the United States. This includes the processing, collating, and fusing capabilities of different organizations and identifies what can be known about the target. This simulation narrows perfect knowledge of the battlefield down to what is really knowable and actionable at specific times in the scenario. Some wargames use human role players to manually narrow perfect knowledge of the real world to what is accessible to our forces. This is effective for small vignettes, but is impractical for large scenarios and long timelines. Neither does it provide the technical accuracy, consistency, and traceability that are necessary for consistent, realistic representation.

Intelligence simulations access the physical battlefield to extract potential targets. They return reports containing the identities, locations, activities, and intentions of enemy forces. This information is used by the Military Simulation to direct its forces and to target precision weapons. This information may also be provided to Political and Financial Simulations to focus their own actions.

Political Simulation

Political Simulations are often conducted as human role-playing exercises. When these are assisted by computers that track decisions and provide information about the world scenario, they can also be integrated into the larger mission against terrorist networks. Actions in the political domain may include garnering support from the United Nations and from countries like Pakistan. They may also impose trade sanctions and withhold foreign aid to induce host countries to take steps against terrorist organizations. The goal is to deny terrorists the safe homeland that they need to maintain training facilities and command bases.

Intelligence simulations provide the data necessary to focus political actions and to convince coalition partners that certain parties are involved in terrorist actions. Political actions should have a direct impact on the ease and strength of operations being conducted in the Military Simulation and may influence the tasking of Intelligence assets.

Financial Simulation

Terrorist networks rely on international money transfers to support their field cells around the world. Political Simulations may provide the evidence necessary to persuade international banks to take actions against specific customers. The political domain is the conduit through which intelligence information is released to the Financial Simulation. The Financial Simulation is where specific aspects of trade and financial sanctions are executed. It also controls the flow of money to field cells, thus impacting their effectiveness.

Financial action against identified groups can lead to information about relationships that identify previously unknown collaborators that can become targets for Military and Intelligence actions. They also provide evidence that can be used to prosecute terrorists in the Legal Simulation.

Legal Simulation

Legal Simulation is a new domain of wargaming. In it, new laws are passed and prosecutions are made against terrorists. These insure that people captured in the Military Simulation and identified in the Financial Simulation are removed from active participation in terrorist activities. This simulation can be used to extract information during plea-bargaining concerning terrorist groups, individuals, and activities that are as yet unknown to our counter-terrorist groups. This information can be fed back into the Military, Financial, Political, and Intelligence simulations to assist them in locating more terrorists.

This simulation may also influence the activities of terrorist groups by intimidating individuals to the point that they do not join the group or are reluctant to carry out missions within the borders of countries that are aggressively capturing and punishing people.

Cultural Simulation

One reason that terrorists can persuade people to join them and to take violent actions against certain targets is that they control the culture in which people are raised. A Cultural Simulation can be used to explore ways to influence this environment. The goal is to persuade these people to take non-violent courses of action to address their grievances. The cultural simulation will represent the influence of media such as the radio and television within a culture. As an example, the Taliban's restrictions against media created an environment in which they could mold people's minds in the absence of contradicting evidence. This simulation would also provide medical and nutritional support to citizens and refugees in target countries. This conveys a distinct message about the supposed enemy—one that contradicts propaganda.

This simulation will have a psychological influence on the people encountered in the Military Simulation. The goal is to influence the populous to support our actions and to sever their ties with terrorist organizations. Assets providing medical support can also gather information useful to the Intelligence simulations in searching for targets or anticipating future activity.

The Gestalt

The federation of simulations described above does not exist at this time. In fact, we are at such a primitive stage in our understanding of all domains of the problem that we are not prepared to begin creating computer software of this problem. At this point we need to approach the problem by manually defining and exercising a wargame that can capture all of the interactions and provide a laboratory for discovering additional resources, events, interactions, and relationships that are important across the entire problem. Then we will be in a position to move on to system dynamics models, operations research, and training systems.

In some cases, the domains and simulations described above are significantly separated along a timeline such that it would be very difficult to include them in an interactive exchange. A scenario spanning several months would be required to pull in all of the assets and events described above.

INTEROPERABILITY ACROSS DOMAINS OF TERRORISM

As stated earlier, all of the activities and interests of a terrorist organization cannot be represented in a single homogeneous model—the problem is too big and too diverse. Such a complex organization is better represented in a suite of models that exchange information that is generated in one dimension of the problem and that also impacts another dimension.

A federation object model (FOM) for such a federation would be very differ-ent from a FOM designed to join simulations of object-on-object combat opera-tions. This FOM would contain many objects and interactions that represent a unique relationship between just two federates. For example, the identifica-tion of terrorist sympathizers may be of no interest to the combat simulation, but may be essential to the financial simulation. This identification would lead financial institutions to specific accounts for confiscation. Financial evidence may then be a crucial part of the legal actions that can be taken against the or-ganization, removing it from the terrorist's support network, and limiting their ability to purchase airplane tickets or to move about the country. Some of the objects and interactions that must exist in a counter-terrorism FOM are shown in Figure 8.4. The figure also illustrates the need for object attributes that are able to respond to influences from these unconventional external federates.

	Physical	Information	Political	Cultural	Financial	Legal
Objects	• Terror Command • Terror Field Cells • Military Forces • Special Forces • Law Enforcement • Customs	• Sensor Platforms • Sensors • Analysts • Distribution Cells	• National Govt • United Nations • Tribal Govt	• Aid Workers • Food • Medicine • Media • Education	• Finance Institutions • Finance Networks • Account Holders • Money	• Justice System • Military Tribunals
Modifiers	• Fear • Loyalty • Duty • Org Visibility	• Cultural Coop • Legal Position	• Power • Attention	• Political Enviro • Cultural Mind	• Self-interest • Asset Access	• Political Positions • Financial Info • Mental State
Interactions	• Prisoners • Communications • Intel Reports	• Tactical Intel • Cultural Intel • Financial Intel • Immigration	• Trade Sanctions • Foreign Aid • Protection	• Lives Saved • Media Images	• Money Transfer • Transfer Source	• Trials/ Convictions • Immigration • Prosecutions

Figure 8.4. Multi-Domain Terrorism FOM Concept

WELCOME TO THE NEXT 20 YEARS

Over the next 20 years, America's arsenal against terrorism will expand dramatically and simulations will be part of that arsenal. Though a need will remain for more traditional models of combat, we are faced with a huge vacuum of tools to address this new threat. We must construct tools that allow our military, political, financial, legal, intelligence, and cultural agencies to understand all of the aspects of this new threat and to become prepared to address it.

Simulation scenarios will no longer focus solely on foreign countries and their topography. In the future, they will include domestic locations and all-to-familiar cultural features.

Chapter 9

SIMULATING DOMESTIC INFRASTRUCTURE PROTECTION

"Al-Qaeda has regrouped and will expand its war to include assassinations and attacks on 'the enemy's weak infrastructure.'"

Abu-Leith al-Libi, Al-Qaeda Spokesman

Al-Qaeda is just one of many organizations that want to disrupt Western governments, businesses, and economies. This decade will require the reorientation of civil and military assets on the terrorist threat. These assets include models and simulations that are used to understand, predict, and rehearse these threats.

The services that tie society together form an infrastructure that spans the entire geography of the country and provides interfaces through which we communicate with the rest of the world. This infrastructure enables the style and standard of living for the country. In many ways, the infrastructure is the physical manifestation of what it means to be a member of the country. Advanced countries are supported by hundreds of distinct and interacting service infrastructures. Protecting these from destruction, disruption, and corruption is a vital part of national security.

National infrastructures are so large, complex, and intertwined that understanding how they work, how they fail, and how best to protect them is a significant problem. Modeling and simulation is one tool that can be used to explore these problems. Simulation tools that capture the behaviors of the systems, the relationships within and between the systems, and that have some ability to measure the impacts of disrupting these systems are an essential part of an effective plan for protecting the infrastructure.

In a previous paper we described the multiple dimensions of the terrorist threat and its relationship to national defense initiatives (Smith, 2002). This paper builds on that work by further exploring the physical domain of the problem, specifically the protection of the national infrastructure (Figure 9.1). The value of applying simulation to this problem has also been recognized within the US government as evidenced by the recent creation of the National Infrastructure Simulation and Analysis Center at Sandia and Los Alamos National Laboratories. Together those labs have a history of modeling large systems such as the energy distribution network, national economy, and city traffic patterns (Nelson, 2002).

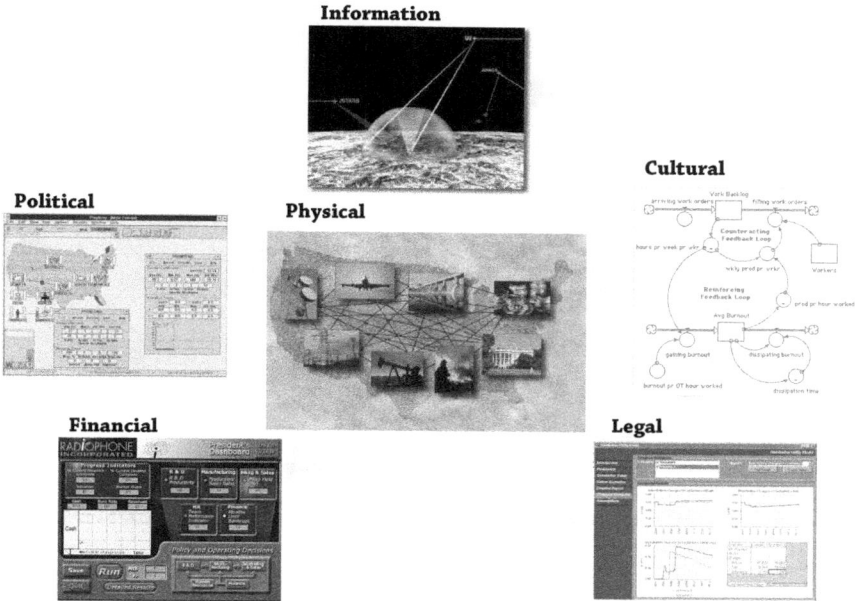

Information

Cultural

Political

Physical

Financial

Legal

Figure 9.1. National Infrastructure in the Physical Domain

We believe that protecting the infrastructure is also a means to another end. The infrastructure is simply a mechanism for providing resources that support the national economy and provide a nation-specific lifestyle (Figure 9.2). The real goal in protecting the infrastructure is to maintain that economy and lifestyle. This change in perspective allows us to search for ways to achieve that higher-level goal. We should consider ways to change the support system for the economy and lifestyles, not just protecting the infrastructure for its own sake.

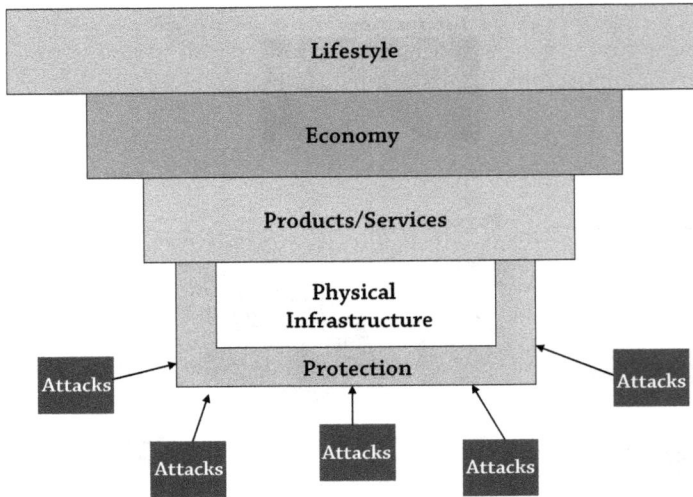

Figure 9.2. Infrastructure Role in Supporting the National Economy and Lifestyle

SIZE OF THE PROBLEM

The term "national infrastructure" is a huge umbrella that covers thousands of different systems and hundreds of millions of users. The most studied of these infrastructures is the electric power generation and distribution system. The second most studied is the telephone system, followed by water processing. All of these are an essential part of the social fabric of a country and constitute assets important enough to protect. However, these are just the beginning of a long list of critical infrastructures that enable a society or country to operate. Table 9.1 lists the infrastructure sectors and systems identified by the President's Commission on National Infrastructure Protection (Marsh, 1997).

Table 9.1. Critical Infrastructure Sectors and Systems

Information & Communications
Telecommunications Internet Public Computers
Physical Distribution
Highways Ports Railroads Waterways Pipelines Air Transportation Mass Transit Trucking Delivery Services
Energy
Electric Power Natural Gas Oil Coal
Banking & Finance
Banks Financial Services Investment Companies Payment Systems Mutual Funds Securities Exchanges Commodities Exchanges
Vital Human Services
Water Sewer Emergency Services (Police, Fire, Medical) Government Services

The fact that so many of these systems are dependent upon electricity makes it clear why electric power generation and distribution is the most studied of these systems. Every one of the systems has experienced outages and the impacts of those outages are understood in a very general way (Robinson, 1998). However, an intentional attack against most of these systems has not been experienced. Neither has an attack against multiple co-dependent systems been experienced. So the cumulative impacts are not clearly understood. Recognizing the immediate effects of losing power to a geographic area for a defined number of days is clear. But, understanding the economic, health, safety, and national security impacts of outages are less clear, especially when multiple outages are experienced at the same time. Simulations can help is understand and explore such events.

We also understand that all of these systems are interrelated. The failure of one network can lead to the failure of another. In the worst case, a domino effect can develop in which multiple system failures are triggered by the initial failure of a single system. As an example, the Federal Railroad Administration estimates that cessation of rail delivery of goods would result in the cessation of automotive, paper, coal, and plastic industries within a few days or weeks (FRA, 1997)

SUSTAINMENT VS. PROTECTION

A simulation that demonstrates the vulnerability of the infrastructure systems listed above is useful in understanding how the system works today. It can be used to identify key nodes that must be protected and key users of the infrastructure that need to be supported by back-up systems and special protection. However, a simulation should also be used to search for better ways to operate and protect the systems and the consumers that depend upon them. The defined purpose of "infrastructure protection" is to erect a defensive barrier around critical resources such that they cannot be penetrated (Marsh, 1997). Given the extreme distribution of these networks, it is unlikely that such a barrier can be constructed. Well-known examples of this problem involve the security of oil pipelines in Alaska and South America. Dissident groups attack these lines at random points across thousands of miles of pipe. Protecting such a distributed network has proven impossible. Oil companies have found that the best solution is to install warning systems that alert the infrastructure owner that a breach has occurred, enabling them to minimize the time to respond to the problem.

Table 9.2. Sample of the Magnitude of the National Infrastructure

137 Major Cities
2,800 Powerplants
   10X Power Sub-stations
463 Skyscrapers
600,000 Bridges
123,000 miles of Railroad Tracks
190,000 miles of Oil & Gas Pipeline
20,000 miles of National Borders

Sources: RAND, FEMA, FRA

Similar issues will exist with all of the systems that make up the national infrastructure, a small portion of which is characterized in Table 9.2. A comprehensive solution should include defense, deception, redundancy, self-healing, alternative services, emergency responses, trained consumers, resource stockpiles, and new expectations from customers. Taken together, all of these actions create a plan for critical service sustainment rather than infrastructure protection. Under this concept, services are sustained by a number of changes to the entire system, such as those shown in Table 9.3. A simulation that can study the different combinations of changes that can be applied and the effectiveness of each of them is a much more valuable tool (BAH, 1997). Such a simulation could play an important role in restructuring and augmenting infrastructure systems such that customers experience a minimal loss of service.

Table 9.3. Critical Services Sustainment

Solution	Description
Defense	Barriers that prevent attackers for accessing or damaging the system
Deception	Decoys that lead attackers to the wrong targets.
Redundancy	Multiple paths and resources for providing services to customers.
Self-healing	Enabling the system to repair itself.
Alternative Services	Providing replacements for primary services.
Emergency Response	Establishing resources and plans for recovering from an attack.
Trained Customers	Teaching the customer how to handle outages and to execute their own recovery plan.
Resource Stockpiles	Identifying the necessary stockpiles to continue operations during an outage.
Modify Expectations	Changing the customer's expectations for service reliability.

Creating a network of sustainment that includes all of the solutions shown in Table 9.3 is a very complex problem. It is more complex than understanding the system as it exists now. Modeling this problem will require the creation of new ways to represent systems. Many of the existing system dynamics tools can combine generic components to create models of specific systems (Robinson, 1998). But, a more comprehensive view of the systems is necessary to represent service sustainment. In such a model, alternative sources of services and the actions of the customer must be represented. Sandia analysts have created generic services modules that takes advantage of commonalities within many of the systems. Their goal is to create a modeling structure and software modules that can capture the interdependencies between each of the pieces by customizing generic models and linking them together to form a complex system.

MODELING SUSTAINMENT

A traditional system model represents a node as an algorithm that responds to stimuli from various sources (Figure 9.3). These algorithms can take the form of a simple equation of growth rates, an index into a table of prepared data, or a reference to an external model. An infrastructure sustainment model should combine contributions from all of the categories listed in Table 2. The infrastructure does not standalone against the attack. Instead a node would represent the infrastructure and its own native ability for self-healing. This would be supplemented with defenses available to that node and the deception that is available to divert the threat. If each of these is defeated, then the degradation of the node would trigger the application of redundant resources such that the customer is minimally aware of the attack. It would also draw upon available emergency response resources to restore operations.

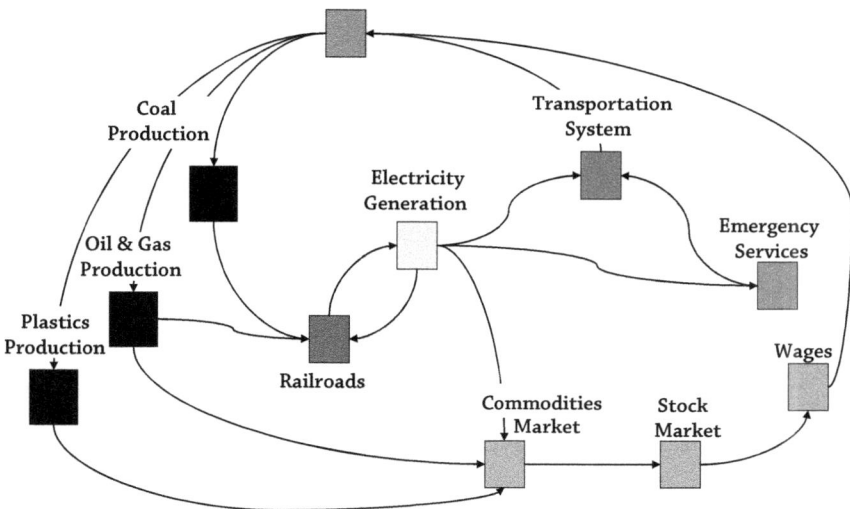

Figure 9.3. System Dynamics Model Examples

In this environment, the customer may experience no loss of services. Or they may need to turn to alternative services to continue their economic operations or maintain their lifestyle. The customer may also require emergency services and may turn to resource stockpiles to continue operations. The model should also include modified customer expectations and an increased reliance on alternatives and stockpiles. Figure 9.4 graphically represents these factors and their relationships to each other.

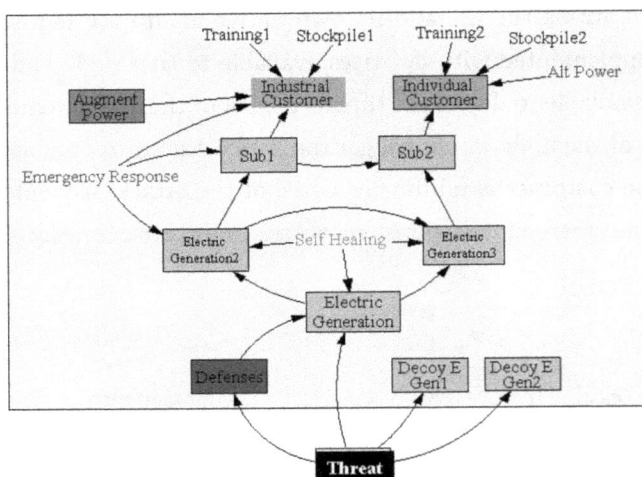

Figure 9.4. Model of Critical Service Sustainment.

MEASURES OF EFFECTIVENESS

A model of all of the cross-dependent systems described in the previous section must include measures of effectiveness (MOEs) that capture the strengths and weaknesses of the aggregated system. The composite system performance may be measured using criteria concerning the performance of the entire system. Some of these MOEs are described in the following sections.

Nodes Down

The number of nodes that are disabled by the attack is an important MOE. It demonstrates both the magnitude of the attack and its effectiveness against protective measures that may have been taken. Since a system can be made inoperable through the destruction of a single key node or by eliminating a number of interrelated nodes, this MOE must include identifications of the nodes affected, their locations, and their degree of connectedness to other nodes and infrastructure sectors.

Time Down

The time that the nodes are not able to provide products or services to their customers is another useful MOE. This measure should focus on the experience of the customer, not on the technical status of the infrastructure nodes. There are many conditions under which the nodes can be considered operational, but the customer still receives no products or services. The important effect is denial of service to the customer, so the health of individual nodes is of less interest than the ability of the entire system to provide services.

Trickle Down

A failure in one system may trigger failure in another. For example, the loss of electrical power can severely impact systems for transportation and communication. So trickledown is a measure of the interdependencies among systems. It may also be important to understand whether a local failure was triggered by trickledown from a single outside system of from the cumulative effects of multiple systems.

Users Down

The number of users that lose service is a good measure of the effectiveness of an attack. This number may be a combination of both direct attack effects and safety shutdowns initiated within the system. This is effectively an internal trickledown effect. The identity and capacity of a specific user is important in measuring the impact of an attack.

Finances Down

Measuring the financial impact of losing pieces of the infrastructure is a very difficult thing to do in a model. There needs to be a way to represent the economic productivity of the systems and their customers. This should be separated into productivity that is recoverable and that is irrecoverable. Late delivery of frozen meat to a grocery store may be a recoverable loss because the meat can be sold the next day. But a day of lost electric or Internet service is not recoverable. In many cases this separation is heavily influenced by the duration of the loss of the infrastructure. Some customers can easily absorb the loss of Internet connectivity for an hour or even a day, but others are not able to tolerate this loss for even one minute.

Confidence Down

Successful attacks will negatively influence the confidence levels of infrastructure customers and of the general populace. This lack of confidence or security can be expressed in many different ways. It may cause people to move away from high-priority targets, to travel less, remain in their jobs longer, take more sick leave, or postpone major purchases. All of these are personal expressions of a change in confidence.

CONCLUSION

Modeling and simulation may be able to provide valuable support to National Infrastructure Protection. One of the most valuable additions would be in exploring alternative modifications to the current methods of protection and service sustainment. The extremely large size of the national infrastructure and the difficulties faced in completely protecting all of the systems involved, make it nearly impossible to prevent attacks against every part of the infrastructure. It is important to recognize that these infrastructures support the national economy and lifestyle and that those are the real assets that we want to protect and that our enemies want to disrupt. Infrastructure protection programs and studies should focus on maintaining the economy and lifestyle, and protecting the various infrastructures across the country is one factor in achieving this.

Chapter 10

SIMULATION IN 60 SECONDS

SIMULATION IS THEATER

"The player must suspend disbelief and forget that it is a game. That shapes and colors the entire experience, gets the player's adrenaline flowing, and makes the whole thing worthwhile."

Jordan Weisman, President of Virtual World Entertainment

When a tank driver climbs into an M1A1 simulator at Fort Knox, Kentucky, he is climbing into a new dimension. He is transported from the training facility onto a battlefield where his life, and those of the entire crew are at stake. This transition happens for a lot of reasons—physical, mental, and sensory. It is important that the simulated driving compartment be small, cramped, dark, smelly, and noisy. All of these allow him to suspend his disbelief and make the mental jump onto the battlefield.

Disney World and Universal Studios spend as much time, money, and attention on the environment around a theme park ride as they do on the virtual experience itself. When a first time customer is walking through a realistic subway station, complete with trains, lights, graffiti, creaking sounds, and oily, metallic smells, that person is being transported from the hot. Humid sidewalks of a Florida theme part, to the cold, busy transportation system of New York City. By the time the customer steps onto the ride, he or she has already been transported to the time, place, and emotional state necessary to make the ride believable.

Simulation is theater and we do our best to transport people to a new environment in the same way that Disney does. Really good simulation will switch on all of the trainee's faculties. It will activate their defense mechanisms, turn on their sweat glands, pump up their adrenaline, and put their mind into high gear.

THE ART SURROUNDS THE AUDIENCE

*"In immersive simulation the audience no longer surrounds the work of art;
the work of art surrounds the audience—just as reality surrounds us."*

Mort Heilig, 1952

In 1952, Mort Heilig had a fantastic new experience. He was sitting in a traditional movie theater viewing a new type of film. To make the film, the crew had mounted a camera on the front car of a roller coaster and made the film from the perspective of the rider. Sitting in the darkened theater, Heilig felt like he was on the roller coaster experiencing the motion of each turn and the sinking feeling in his stomach when the car dropped down the side of a steep hill. This experience taught him what it really meant to immerse an audience or user in a simulated environment.

Heilig drew a distinct line between art that the audience looks at and art that the audience is a part of. Immersive simulation is a form of art that surrounds the audience. It consumes them the way the giant fish swallowed Noah. The audience sees the work from the inside as a participant in the action, rather than from the outside as an observer of the action.

SIMULATION MUST FIND ITS OWN FORM

Literal representation of the world is the wrong approach. Just as artists interpret the world to more strongly express their message, a simulation must find a form that expresses its unique message.

Every simulation system is different. Each captures a unique subset of the real world with unique rules and relationships for manipulating that subset. Like snowflakes, no two simulations are the same. Therefore, the visual, textual, and data windows into the virtual world should be unique for every simulation.

Experienced simulation professionals and familiar with the typical 3D virtual world and 2D map display as methods of expressing the content and activities of the simulation. But, a simulation that represents aggregate military units cannot express itself with the same icons, maps, rules, and tables that are used to communicate logistical models or intelligence collection.

Simulation science and simulation projects need to constantly seek out newer and better ways of communicating with the users. The 2D maps that have evolved over centuries contain the natural limitations inherent in paper maps. Simulations are dynamic, ever changing machines. They need a more fluid, multi-dimensional style to communicate the richness of activity that is occurring within.

THE TWO FACES OF SIMULATION

Virtual simulation challenges your reaction times by immersing you in a dangerous situation. Constructive simulation challenges your strategic mental abilities by forcing you to solve a complex problem involving other people and assets.

First Person, immersive simulation is a beautiful and exciting experience. The three-dimensional images, beautiful textures, synchronized sounds, and real-time reaction to your inputs makes it a thrill to be connected to. This is the flash that hooks an audience, draws them in, and keeps them working with the simulation. The computer game industry has harnessed this immersive environment and made it even more addictive. They have removed many of the complex and tedious tasks that were part of the industrial and military training environments and turned the virtual world into pure play.

However, first person immersive experience is not the only important or reward role of simulations. Third Person, objective simulation is a thoughtful learning experience that provides depth and a rich environment that can be explored in more detail. The lessons to be learned and secrets to be discovered are less obvious and involve many more factors that most immersive simulations. These worlds challenge your mind to find options and solutions that involve a large number of variables. They seek to teach lessons that are more complex and nearly impossible to convey in less interactive forms like textbooks, classrooms, and maps.

Both of these faces offer important contributions to training, teaching, and molding people to do their jobs better. Jobs that are primarily based on personal reactions to the environment require first person immersion. Jobs that are strategic thinking require a third person, leadership perspective that encourages creative mental consideration. Jobs that focus on team coopera-tion may fall into either category, depending upon the role of each person in the team. A team of soldiers operating the machinery inside of a tank may be best training in an immersive environment. But, a command team that is re-sponsible for hundreds of other people and assets at remote locations may be best trained with a third person, leadership system.

FROM WHENCE SIMULATION

Strategic thinking tools emerged in the form of games as early as 3000BC. These games spread through China, Japan, India, and Europe for 3500 years before creating the game of chess that is so widely recognized today. Virtual simulations emerged directly from live, non-lethal exercises and martial practices. The motivation to simulate is ancient—computers have just given us a powerful new tool to satisfy this motivation.

Simulation games emerge around 3000BC, during the Bronze Age, with the invention of Wei Hai in China. The name means "encirclement". Unfortu-nately, no remnants of the game remain, so we can only conjecture that is was something like the more familiar Go in Japan.

Wei Hai is believed to be the predecessor to Japanese Go, which emerged around 2300BC. This led to the invention or emergence of Chaturanga in India around 500BC and Chess in Europe around 500AD.

Military board games were invented in Germany in 1664 and evolved over the next several centuries. These tools provided both entertainment and train-ing in strategic thinking. Most of them require that the player plan for the good of a number of pieces and the supporting synergies that can be developed between many of them.

Immersive computerized environments are a direct extension of live training. The drills of Roman legions, stylistic defenses in the martial arts, aircraft combat practice, and rifle ranges all influenced the forms of computer simulators that were built from the 1930's through the present day.

In the early 1980's the Atari arcade game Battlezone was the inspiration for a $500 million military project to create globally distributed tank team trainers known as SIMulator NETworking (SIMNET).

ANATOMY OF A SIMULATION

Military simulation systems consistently include six basic components.

- The Simulation Engine contains the calculations that make-up the model of the virtual world. This virtual world includes objects, physical and cognitive behaviors, and an environment.
- A Training Interface allows the player to view, manipulate, and participate in the virtual world.
- A Controller Interface creates "the man behind the curtain" that makes the simulation operate for the stimulation and entertainment of the player.
- A Scenario Generator allows the operator to create the objects and environment of the virtual world.
- The Analysis Station collects information about the execution of the simulation and the player's experience, replaying it for study and instruction.
- The Network Interface allows the virtual world to exist and interact with multiple players across a computer network.

THE PRINCIPLE OF THE THING

Every simulation system incorporates a set of common principles.

- It must understand and meet a user's needs,
- It must exist at an appropriate level of detail,
- It must build upon the lessons learned from previous simulations,
- The ideas must be explored with a prototype of the final system,
- The results must be credible to the user of the system,
- The simulation must use a valid set of data,
- The data and the software must be separated to allow the expansion and modification of the simulation,
- It is subject to the universal constraints of cost, schedule, quality, and competent staff.

1,000 WAYS TO MISS THE TARGET

Every simulation is a creation from the mind of a group of developers. This creation may or may not accurately represent the world needed by the users.

Echoing the concepts put forward by other experts in modeling and simulation, Phil Kiviat explains that a model only has to be as good as its intended use. This parallels the ideas of limiting the details incorporated into the model that were stated by Robert Shannon and Alan Pritsker.

Every model is a tool that has a specific purpose. Some models have a single function, like a screwdriver, others have multiple functions, like a Swiss Army Knife. But every one of them is limited in the type of problems that it can be used to solve. The model should be designed to solve these problems in the most efficient and accurate manner possible. No model is an attempt to solve every problem or answer every question in a specific area.

WHAT IS THIS GOOD FOR ANYWAY?

Simulation presents many advantages over experimentation or training with real systems in the real world. These are good reasons that simulation has been successful and continues to grow.

Simulation is a cost effective means of learning lessons and training people. It allows the controllers ultimate situation control and presents a must safer operational environment. With simulation it is possible to generate all conditions of interest at will. The environment does not impeded experimentation or training conditions. Finally, simulations do not disturb or destroy the environment in which they are conducted.

WHAT IS THIS NOT GOOD FOR AFTER ALL?

"Be skeptical about the value of modeling and force the 'why do we need it' discussion at the start of the project."

John Sterman, MIT

As professional simulation and modelers, we are often hesitant to question the need for our expertise. But, it is essential that you establish why modeling is right for this project and what you are expected to contribute right up front. In some cases you may discover that additional testing, a better design process, or formal discussions with the users are sufficient for the problem at hand. In other cases these questions will allow you focus on solving the right problem, establish a realistic schedule, and set customer expectations at the right level.

Chapter 11

STRATEGIC DIRECTIONS FOR DISTRIBUTED SIMULATION

This chapter presents strategic directions and research challenges in distributed simulation. In searching for these technologies the author polled several prominent members of the simulation community and reviewed recent publications that characterized many areas of simulation:

- *Proceedings of the 1998 Winter Simulation Conference* (Medeiros, Watson, Carson, & Manivannan, 1998),
- *Proceedings of the Twelfth Workshop on Parallel and Distributed Simulation* (Unger and Ferscha, 1998),
- *Proceedings of the 1999 Game Developers Conference* (Yu, 1999), and
- *Digital Illusions: Entertaining the Future with High Technology* (Dodsworth, 1998).

Distributed simulations are those applications that span multiple computer devices, executables, or geographic areas. These include what are often referred to as parallel and distributed simulations (PADS) and distributed interactive simulations (DIS) (Fujimoto, 1998). These communities vary widely in their techniques for implementing a distributed simulation, but they both fall under the general category of distributed simulation.

Distributed simulation is widely applied in military training systems in which computers and executables have been joined together through techniques like the Distributed Interactive Simulation (DIS) protocol, Aggregate Level Simulation Protocol (ALSP), and the High Level Architecture (HLA). It is also used in analytical models in which networked and parallel computers divide a problem into smaller pieces that can be solved more rapidly. The entertainment community has applied distributed simulation ideas in attractions like the Battle Tech Entertainment Center and the Internet-based Virtual Worlds environment. Most computer games also contain a distributed simulation mode that allows them to interoperate with other people playing the game on the Internet. Games like Quake II, Rainbow Six, Command & Conquer, and the entire Star Wars line are well know and well sold for this capability.

STRATEGIC DIRECTIONS

The strategic directions are areas in which simulation can be applied immediately, but where we have not taken full advantage of the technology that is available. These include:

- Systems management,
- Real-time decision making,
- Persistent virtual worlds, and
- Virtual verisimilitude.

Systems Management

It is possible to embed simulation modes in the operating systems of computer systems. These systems can feed data about their operations into a data store that is accessible to simulation processes. Periodic execution of these would evaluate this performance data and identify the operational trends in the data. This can then be used to optimize the system for its most characteristic applications.

The PC is a general-purpose computer that is put to specific tasks once it is in the hands of the user. If the operating system contained a simulation kernel it would be able to evaluate the uses to which each machine was being put and optimize that machine for those applications. The simulation would need a database of application characteristics such as word processing, accounting, databases, graphic art, sound editing, games, web serving, web surfing, telephone management, and hundreds of others.

The advantage of simulation-based adaptation is that the user need not be an expert in configuring the machine and the simulation can re-optimize the machine when it is applied to a different function. Since most systems are used for more than one application, the simulation would also be able to adjust the configuration to best satisfy two or three applications—a task beyond the abilities of most PC users.

Real-time Decision Making

The world is filled with opportunities to apply computer simulations to assist in real-time decision making. Any place that information is available in a digital form and humans are evaluating that information to making decisions based on that information, there is an opportunity to support the human with a simulation.

These opportunities occur in thousands of fields, only a few of which will be described here.

Combat Consultant. Large military organizations are migrating their communication and decision-making tools to computer systems. This provides combat information in a form that can be accessed and evaluated by a simulation. We are lucky to live in an age in which our citizens are not faced with life-and-death combat decisions on a daily basis. As a result, soldiers that encounter this kind of event are relatively inexperienced at dealing with it. Military organizations mitigate this through extensive training activities (some of which also involve simulations), but there is no substitute for experience. A Combat Consultant is a simulation mode embedded inside of the command, control, communications, computers, and intelligence systems being used by the soldiers (these are commonly referred to as C4I Systems). The simulation is equipped with the expertise of previous commanders and the real-time expertise of other commanders currently using similar systems on the network. The Combat Consultant can monitor the information in the system and suggest alternatives that may be successful under the current situation.

Though this may begin as an expert system, it also includes the real-time experience of other commanders solving similar problems at this moment in time and a simulation engine to project this situation into the future. The system searches for the best strategies for handling each combat situation in real time.

The term C4I evolved from C2 over the last two decades to more accurately describe the operations performed by commanders and their decision support systems. It is time to add simulation to the acronym—C4IS.

Aircraft Navigation. The Federal Aviation Administration is planning to change the mechanism for controlling commercial air traffic across the country. Under the new method, entitled "Free Flight", pilots will have unprecedented decision making authority in selecting their flight paths and adjusting them throughout the trip. Simulations can assist these pilots by evaluating environmental data, aircraft status, data received from sensors, and data transmitted from the ground. The simulation can constantly study the current situation, looking for the optimum solution for reaching a destination. Perhaps more importantly, the simulation can also generate customized plans for use in an emergency. When the unexpected happens, a plan is ready and the flight consultant is there to support a pilot who is confused, scared, and unable to make decisions.

Crowd Management. All large cities face the problem of managing the flow of people trying to accomplish their own objectives. These people may be driving in rush hour traffic, searching the mall for a sale, or rushing to the best rides in a theme park. In all of these cases, we could optimize operations by directing this traffic. Using traffic flow sensors we can measure the location and density of people in the system, feed this information to a simulation, and look for solutions in real time.

In the case of the theme park, entertainment events could be scheduled by the simulation in patterns that push and pull the guests in specific directions. Good theme parks are designed to direct the flow of traffic from the time you enter the main gates until you finish your tour of the attractions. These designs would be assisted in real-time by simulations that recognize overcrowding in one area and schedule activities to pull part of the crowd to another area. The "pull" mechanism may be the appearance of a costumed character down a side path; beginning a computerized entertainer directly behind the accumulating mass; or the sounds of a roaring dinosaur in a different direction. These tactics are designed to redirect the crowd in a manner that is non-intrusive and that appears to be of the guest's own volition. Events may also be scheduled to direct the guest's attention away from the fact that they are waiting in line.

Market Prediction. Banks and financial institutions are already using simulation and gaming techniques to analyze past performance and predict future activities. These simulations influence commodity trades, stock speculation, and currency exchanges. They provide an edge over competitors that can result in millions of dollars in additional profits. Simulations of this type can be embedded into many forms of stock selection and advice software, including those used by your stock broker, internet stock trading web page, and personal asset management package (e.g. MS Excel, Quicken, MS Money). These are also useful tools for teaching a novice how markets work and what to watch for in future investments.

Persistent Virtual Worlds

The networked world is a natural host for a persistent virtual world that is accessible to all users. We need to create virtual environments that are persistent over many years and that form the foundation for specific studies, training, and entertainment that will be conducted within them. The gaming community has already accomplished this with online persistent virtual worlds like Ultima Online, Diablo, and Everquest. These provide persistent fantasy worlds that evolve as the users interact with them.

Similar virtual worlds need to be created by high level sponsors of studies and training events. These would be the seeds from which scenarios are drawn and the environment in which distributed interactions occur. Organization like the Office of the Secretary of Defense, the Defense Modeling and Simulation Office, the Central Intelligence Agency, the National Air and Space Administration, and others need to become the hosts for persistent virtual worlds that support the needs of their entire customer base. It should be possible for globally distributed customers to enter these worlds at any time and explore solutions to current problems.

Commercial versions of this can be used to track the activities of specific individuals in the population. The popularization of cell phones and pagers has placed electronic tracking devices on the belts of a demographic of people that we are most interested in tracking. These tagged people can serve as a sample of the general population, allowing us to see customer density in airports, malls, highways, and large entertainment events.

This could be used to identify potential witnesses to crimes based on their presence in the area and predictions of the path they were likely to have followed while in the area. It may even be possible to identify the perpetrator of the crime using this technique.

Virtual Verisimilitude

In the simulation business we strive to create virtual worlds that accurately represent the real world. This always involves a high degree of abstraction to help us experiment with systems that are far too detailed to fit into any model. However, we have been so conditioned by our lack of computational power and seduced by our skills at abstraction that we sometimes avoid extending our simulations when we have the tools to do so.

There are few simulations that portray a really convincing virtual world. With all of the computational power now available and the increasing maturity of software tools to build models and virtual worlds, we need to explore a new level of representation. It is time for the next big advance in modeling detail and the richness of virtual environments

Statistically accurate simulations are excellent for many applications, but we need to begin equipping ourselves with models that accurately represent individual objects, events, and interactions without relying on actuarial effects to make them correct.

RESEARCH CHALLENGES

The research challenges are those technologies that are essential for the progress of the field of distributed simulation. Though there are many areas of valuable research, the four listed here are broad enough and essential enough to be listed as research challenges. These are:

- Human behavior modeling,
- Simulation domain architectures,
- Abundant network bandwidth, and
- Practical event management techniques.

Human Behavior Modeling

Many simulations are driven by statistical distributions that characterize the average behavior of a system, but do not claim accuracy for individual events or small time intervals. These distributions represent the activities of machinery, the population growth rates of animals, and human performance of specific tasks. However, they do not model instantaneous behaviors of intelligent or reactive beings in the virtual world. We are in dire need of techniques for inserting intelligent, reactive, unique human behavior in the virtual world.

Military training simulations and computer games require interactions between human operators and automated virtual humans. In the past, this has been accomplished through techniques like Finite State Machines that encode specific behaviors and define the transition conditions from one behavior to another. However, we are discovering the limitations imposed by this technique. These are very difficult systems to create and maintain. Human operators that interact with them regularly identify their limitations and take advantage of them. The entities controlled by these techniques do not exhibit realistic behaviors, rather they exhibit correct behavior—"by-the-book", robotic actions.

We need to discover and create techniques for representing the behavior of human leaders, followers, and groups that give them the ability to appear "live" or "real" to the humans interacting with them. Both the military and the gaming communities are augmenting their robotic methods with "softer" models that include human emotion, training, and fatigue. These result in objects that are all slightly different in spite of being driven by the same software.

The distributed simulation community needs a set of behavior libraries that can be linked into a simulation in the same way we currently link in statistical distributions. This will require the definition of a set of categories of behavior and API's that are necessary to stimulate those categories.

Domain Architectures

Within the Department of Defense we have been developing standard protocols for joining multiple, previously independent, simulations. These methods have included the DIS protocol, the Generic Data System (GDS), ALSP, and most recently, the HLA. With HLA we have begun to identify simulation functionality that is generally necessary for all systems and which should be included in an infrastructure rather than within specific simulation models. This approach allows a simulation development team to reuse some of the essential functionality that is included in the general infrastructure. It has also encouraged us to question the uniqueness of every simulation system. We recognize that simulations fall into domain areas in which the degree of commonality is much higher than it is across all simulation systems. We begin to imagine a layered view of simulation uniqueness similar to network protocol layers. Higher layers become more specific until they narrow to a single application.

It should be possible to develop an architecture that supports an entire domain of simulation systems, providing a large common pool of functionality. These architectures may include a general interoperability standard like the HLA, but would go further by defining a set of domain tools for operating the simulations, common interfaces to connect to external systems, and object base-classes from which to extend unique object instances.

Information Bandwidth

Distributed simulations can not exist without sufficient reliable communications bandwidth for delivering events and synchronizing execution of the entire system. This bandwidth is currently one of the limiting factors on the size of a distributed simulation. Luckily, bandwidth is also a limiting factor for all applications using the Internet. This has attracted millions of commercial research and development dollars to the problem. That work can and will be applied directly to simulation applications. The global communications industry will discover methods for providing abundant information transfer. These will include methods for configuring the physical medium of delivery and efficient protocols for transferring data. We may productively put our efforts into simulation-specific communications protocols that are not addressed by other communities.

PDES Management

For twenty years we have been involved in research to discover techniques for practical and efficient synchronization of distributed simulation processes. This has resulted in some very clever and powerful ideas for addressing this problem. However, these ideas have been embraced by few industrial and government applications. The constructive wargaming community has adopted Conservative Time Management, but Optimistic Time Management is still searching for an ideal application.

We must identify applications that are well served by the different methods of event management. To justify further study, our research in this area needs to find a practical and valuable home in commercial, government, or military simulation systems. By 2010 we should be able to apply the appropriate synchronization technique to a distributed simulation by analyzing the problem, setting configuration variables, and attaching the event management engine to our simulation. Trial-and-error and fine-tuning of the engine must become standardized such that a simulation professional can perform these operations, rather than a PDES specialist.

CONCLUSION

The strategic directions and research challenges presented in this paper emphasize two different aspects of the future of distributed simulation. The first is the need for additional development and imagination in applying the technologies we already have. The second is the need for additional research and innovation in areas that will allow us to advance the state-of-the-art. It is the author's opinion that, while the research challenges provide stimulating problems, the strategic directions for applications are much more urgent at this time. The world is in the middle of an information, communication, and computational explosion. Thousands of advanced applications are being fielded every year and many of these could be improved through the inclusion of existing simulation technologies. However, these opportunities are being lost or the technology reinvented by others because of the lack of communication, marketing, and proselytization by members of the "core" simulation community.

Chapter 12

THE FUTURE OF VIRTUAL ENVIRONMENT TRAINING

There is a rich history in researching and developing virtual environments within the military. The Simulator Networking (SIMNET) program of the late 1980's and early 1990's demonstrated the deep value of virtual environment applications (Miller and Thorpe, 1995; Davis, 1995; Singhal and Zyda, 1999). Twenty-five years later there have been significant advances in this area, but there remains vast unexplored potential in this field. There are potentially hundreds of valuable applications to real military operations in logistics, command and control, situation understanding, and information fusion. In both the commercial and the military worlds, the power of VE is significantly enhanced by the growing availability of digital data in every industrial and government domain. In a world where reconnaissance photos are captured on physical film, there is little that computation and VE can do to enhance this information. Once those photos become digital, it is possible to analyze, fuse, integrate, and morph them so that they become the visible skin of a VE. As most information about the world becomes digital, it creates opportunities to generate higher levels of understanding and new advantages over competitors. As the world has become networked, digital data has also become globally accessible so that digital photos from every continent can be viewed in real-time anywhere in the world. As network bandwidth, computational power, and VE algorithms advance, there will be a point at which these images can be stitched together into a seamless three dimensional map of the entire world and navigated in real-time. This data will include digital images, sound waves, weather patterns, population densities, personal locations, RF spectrum, financial transactions, and dozens of other specializations.

From a military perspective, most situations of interest are geographically based. In the past, our technologies have limited our ability to construct information into a geographic form similar to the world from which it was collected. The VE is a new and powerful alternative to textual, graphic, and other paper-oriented representations that have dominated our decision making for centuries. Today's leaders, managers, and engineers are very comfortable communicating information that has been structured in the form of graphs and tables. The next generation will be just as comfortable structuring information into unique VEs and exploring those collaboratively as a means of understanding and manipulating the world.

COMMERCIAL LEADERSHIP

Sometime in the late 1980's there was a tipping point (Gladwell, 2002) at which commercial industry took the lead from government laboratories in advancing computer technologies. The explosion of consumer-grade computing power led to a corresponding explosion in software applications that could exploit this power. One of these growth areas was the computer gaming world that created products like Quake, Unreal, and an annual harvest of new competitors presenting the best VE rendering available at consumer price levels. This civilian market will continue to drive research and development into VEs and the creation of ever more beautiful and immersive worlds in which to interact with information and other people (Smith, 2006; Dodsworth, 1998).

Just as e-mail and instant messaging have replaced the telephone as the leading medium for personal communication and the Web has replaced the library as the leading repository of information, VEs will replace the textual Web page as the primary medium for shopping, socialization, and exploration. VEs can capture both the contextual relationships of hyperlinks and the proximity relationships of geographic collocation. Some form of VE will become the context within which online digital information is organized, significantly extending the linked, flat web pages that convey this information today. People who are browsing through data will be able to discover related items that are geographically close to each other just as they do when browsing in a physical library or bookstore. Applications like Google Earth, Second Life, and World Wind are beginning to illustrate this future. Imagine a World Wide Web in which all personal information is tied together in a single context. For example, a social network of friends live as 3D avatars in a VE apartment where favorite video clips are streaming on one wall and the contents of an online encyclopedia are lying on a coffee table. Further, in a VE there is no reason that the apartment has to look anything like a physical habitat; it could be a giant garden, forest, cloud city, or ant colony. The information that people need and enjoy may grow like flowers in the garden all around them, their colors and sizes representing currency, importance, source, or other key attributes.

Most commercial VE expressions are uniquely personal, playful, and civilian, but the technologies behind them are seriously powerful. Like the radio and the semiconductor before them, these technologies are not limited to entertainment, business, or national defense, but can be applied equally to each domain. The commercial world will be the source from which advanced VE technologies spring and the foundation from which military applications are built.

Though computer generated VEs are primarily visual, there may be other alternatives to loading information into the human mind. Direct neural stimulation may allow information to enter without going through the eyes. Technology that enables a blind person's mind to "see" is similar to that required to generate a VE directly in the mind. The advantages of this approach are beyond current understanding. A neural image may be superior to a standard visual scene. It may create a new sense of the data that is contained in the world, effectively enhancing the human ability to perceive rich mixtures of data within a VE.

Further a field is the possibility of creating or enhancing the VE through the use of chemicals. It may be possible to chemically stimulate the brain to construct useful representations of information. The 1960's experiments with LSD cast a dark shadow over these kinds of experiments, but new research into chemically enhancing athletic and soldier performance are bringing these ideas back into vogue. Just as caffeine can enhance alertness and reaction time, other chemicals may improve understanding of information that is part of combat operations or that drives training for life threatening missions.

VE APPLICATIONS

The term "serious games" is often used to describe the application of game technologies to military or industrial problems. This has been a useful term, but it will become archaic as the distinction between game technology and non-game applications fades away. Computer chips and graphics cards are not referred to as "entertainment chips" or "serious graphics cards." They are just tools for constructing useful applications. The same will occur with "serious games." All industries will have VEs that meet their needs, just as they have specialized computing and communications devices today (Bergeron, 2006; Lenoir, 2003).

Since 1992, the military has identified its simulation tools as Live, Virtual, or Constructive. This delineation has highlighted the computational and conceptual limitations in representing both breadth and depth in a VE. "Virtual" refers to the use of simulated objects by real humans and these systems have typically represented small areas with few objects at relatively higher levels of detail. "Constructive" refers to the use of simulated objects by simulated people and these systems have represented very large areas and many objects with relatively less detail. In the years since these definitions were standardized, advances in computation have enabled the creation of many systems that combine one or more domains. Further advances in computation, communication, and conceptualization will allow us to stretch the boundaries of these domains so that there is little difference between them. In the future, "Constructive" and "Virtual" will refer only to the view that is being presented to the human or to an AI, and not to any inherent limitation of the models that are driving the virtual world.

There have been three distinct generations of "Constructive simulation" and perhaps future VEs will create a fourth. The first was the use of sand tables and miniature figures, essentially a scaled representation of the real battlefield. The second was the paper board game that allowed greater abstraction and additional rigor in the rules and mechanics of behavior. The third was the computerization of the wargame which extended the algorithms to the limits of the computer, rather than the limits of a human player (Allen, 1989; Perla, 1990; Dunnigan, 1993). Advances in VE will enable the creation of a constructive simulation that is just as detailed as any virtual simulator if so desired. It will employ aggregation and abstraction as a useful metaphor, rather than as a core design limitation driven by limited computational power.

In the "Live" domain there will be VEs embedded in real equipment just as two-dimensional map displays exist in equipment today. These VEs will be integrated into the control screens and head-mounted displays that are currently portals into flat, disassociated, two-dimensional data. Rather than seeing the battlefield from a top-down, two-dimensional view, the operators will be able to see it in three dimensions from any angle that they find useful. This is a hugely powerful paradigm and carries so many potential options that the challenge will be in determining where the valuable views lie, not in rendering and animating them for the operator. In this world, there will be little difference between the objects that come from a simulation and those that exist in the physical world. All of them will be seamlessly integrated into a VE.

ARMY MISSIONS

VEs are supplemented with physical and cognitive models, software management and control tools, and external interfaces to operational devices to create simulation-based training systems. As the nature of the Army mission has changed, simulations and VEs have been challenged to represent new missions, new threats, and new tactics that capture the essential elements of the real world and can be used to teach this reality to humans. We have emerged from four decades of a Cold War in which most military training has focused on large combat operations that occur on specified battlefields where all participants were expected to be combatants. More recent missions have focused on small units in an urban environment where they must perform humanitarian operations, search and reconnaissance, facility defense, and combat operations all on the same day. This has created a situation in which our VEs and simulations are expected to represent a much more diverse set of objects and interactions. These can no longer be "combat only" models of the world. The focus of current and future missions appears to be on much smaller areas, making it both possible and desirable to deliver very high levels of detail in the area of operations. This detail calls for a VE that can recreate combat operations in a single city block, but also allow personal communications with the populace to build an understanding of the societal factors surrounding the military operations. These factors will trigger important actions and reactions as the simulation progresses. Many of the current simulation models focused on immediate action and immediate consequences. In most cases, these actions/consequences are discrete and do not influence actions between objects in the future.

While the military simulation community has been wrestling with models of information processing and human reaction, it has just begun to explore the richness of person-to-person relationships and their influence over different groups within a population. There is a great deal of "soft social science" that needs to be incorporated into VEs in the future. Accurate physics models of weapon penetration and aircraft lift remain important, but a useful understanding of the urban battlefield is driven by human interactions, motives, and group dynamics. In the past military simulation systems have been able to focus on the universal and verifiable behavior of the physical world. But models of personal relationships and group behavior are highly cultural, social, and geographical. Huntington (1996) has suggested that all future competitions will be based on seven unique cultures that have emerged in the world, Western, Orthodox, Latin American, Muslim, Hindu, Sinic, and Japanese. Rather than a bi-polar world threatened with traditional combat, we live in a more complex world in which the confrontations may be focused in the political, military, economic, social, infrastructure, or information domains and involve seven different and powerful cultures. VEs that are able to represent such a diverse world accurately and effectively will be a significant challenge and a significant focus in the future.

ADVANTAGES

VEs that are created electronically, biologically, or chemically all present significant advantages for military operations and training. They create an improved space for accessing, absorbing, understanding, and applying information. These are all information-based terms that create a pattern very similar to the Observe, Orient, Decide, Act (OODA) Loop that was first proposed by Colonel John Boyd (Coram, 2004).

The advantages to be gained are so significant that VEs will continue to grow in importance and in the breadth of their application. Specialized versions of VEs will be used for hundreds of different applications, each with a unique focus, but built on a core set of technologies. As the limitations of computer and communication technology fall away and our level of expertise in creating and manipulating these environments increases, VEs will appear in all types of consumer and military systems to aid people in making better decisions and taking more appropriate actions. VEs combine technologies that have been maturing in the training, entertainment, computer science, and communications domains for several years and have reached a point at which they can be adopted by hundreds of commercial and government organizations.

Part 2

SERIOUS GAME
TECHNOLOGY

Chapter 13

DOES GAME TECHNOLOGY MATTER?

Among the ruins of ancient Egypt there are multiple references to games that were popular among the Pharaohs. The remains and images of the game of Senet date back to 3,000BC. This board game contains features similar to modern checkers and a method of play reminiscent of a horse race around the board. Though primarily a game for entertainment, it was also used as a mystic tool to foretell the future. Egyptians believed that the square that a player's piece ended on contained special significance about what would happen to the person in the future. Though we would consider this superstition, the players at that time took the results as guidance on decisions about commerce, farming, religion, or family.

Around 1,400BC the game of Mancala emerged in Africa. It was a tool used to account for livestock and crops, and a form of entertainment. Tribes-men used the board and stones to negotiate the trade of goods, and perhaps to gamble for a better exchange. But they also passed the time in the fields playing a version of Mancala that had no economic consequences, but was purely a form of entertainment.

In 1956, Charles Roberts developed the components of the modern board wargame as a tool to help him prepare for his commission-ing in the U.S. Army. But by 1958 he realized the commercial value of this wargame and cre-ated the Avalon Hill game company to market it to thousands of avid "arm-chair generals" who were eager to test and develop their own tactical military skills, but for entertainment. For the next four decades Avalon Hill and several competitors created wargames for both entertain-ment and military training.

Were these games primarily and initially enter-tainment or serious tools for guiding life decisions? There was really no hard division between the two purposes. There is no law of nature that says tools for education and training cannot be enjoyable to use, or that such tools cannot be inspired by or created from applications that were initially entertainment. The dual nature of games has been with us for at least 5,000 years. Today we may have replaced dice made from sheep knucklebones for computerized, pseudo-random number generation algorithms, but we con-tinue to look to the results of game play for insight into important problems in our lives. Now we place our faith in the accuracy of mathematical and logical algorithms rather than the mystical forces influencing the roll of the die, but we continue to construct games that can challenge our thinking and guide us to a better understanding of the world.

WHAT IS A GAME?

What makes some activities and tools into games, while others are considered completely serious tools? In his 1970 book entitled *Serious Games*, Clark Abt defined a game with these words, "reduced to its formal essence, a game is an activity among two or more independent decision-makers seeking to achieve their objectives in some limiting context. A more conventional definition would say that a game is a context with rules among adversaries trying to win objectives." In a 2005 issue of *IEEE Computer*, Mike Zyda defined a game as, *"a physical or mental contest, played according to specific rules, with the goal of amusing or rewarding the participant."* He went on to suggest that a serious game was, *"a mental contest, played with a computer in accordance with specific rules that uses entertainment to further government or corporate training, education, health, public policy, and strategic communication objectives."* Zyda explicitly points to the desirable goal of using "entertainment" to further the goals of the organization, to harness entertainment, fun, engagement, challenge, and trail-and-error to get people to learn more or to learn faster.

Academics like Andrew Hargadon at University of Southern California explore the difficulties involved in adopting tools and practices from other industries. There is a psychological, social, and professional barrier that keeps people from accepting ideas that were "not invented here." The barrier between "serious business" and "frivolous entertainment" is even higher, wider, and deeper than those between industrial professions. Industries may adopt new computers, networks, materials, and energy sources. But reaching into the entertainment industry for something that can improve effectiveness is considered quite a daring and questionable move.

GAME TECHNOLOGY

Games have created and introduced new technologies for centuries. Ancient games offered numbered throwing sticks, the predecessors to dice and random number generators, as a means of making decisions with limited information. Board wargames of the 1950's introduced the hexagonal tessellation of terrain, a concept that is still used in cellular communications models as an approximation to the circular area covered by a tower. Charles Roberts introduced the combat results table as a means of enriching the military results from the throw of a die. Today all military models use extensive algorithms to make decisions, but often retain a random number generator as a nondeterministic influence in those algorithms.

Currently it is difficult to determine whether computer hardware and software technologies are "game technologies" or "serious technologies". Graphics cards, network cards, and multi-core chips are all essential for the play of the latest computer games. But should they be tagged as serious or entertainment technologies? Does it matter? Does it help?

Recently the gaming industry has been the source of some of the best software technologies on the market. The 3D scene generators or game engines are far superior in performance and features to competing applications created in serious industries and academia. Game companies have adopted the principles of man-machine interfaces and effective graphical user interfaces to create complex applications for which no user's manual is required. But similar interfaces in serious industries can be so complex that multi-day courses are required to learn to use them. Games have isolated the most essential physics and human behavior features such that they can be incorporated into an application that can run on a consumer PC. They are certainly not the highest fidelity models of physics or artificial intelligence, but they are the most accessible and among most useful. Multiplayer games have advanced networking protocols and libraries so that players can join the virtual world from anyplace on the planet. But what serious industry applications provide this type of ad hoc collaboration?

The financial incentives and the personal energy that drive the creation of new technologies in the game industry have led to technologies that are just too valuable to be excluded from other serious industrial applications. All industries have got to take these technologies seriously or risk being passed by competitors who will use them.

DOES GAME TECHNOLOGY MATTER?

Game technologies have been adopted for military training, medical education, emergency management, city planning, spacecraft engineering, architectural design, religious proselyzation, political communication, movie making, and advertising—to name a few. These are far from being the dominant applications in any of these fields. But they gain ground every year as young game players become serious business people and as older business people be-

Edison Electric Company

come more avid game players. The barriers are falling. Each year more people are able to peer through the science fiction veneer of a space game and see the powerful computer science beneath. They understand the advantages of putting this technology to use, and doing so before a competitor does the same. In his 2003 *Harvard Business Review* article entitled "IT Doesn't Matter", Nicholas Carr shook up the business and the IT worlds with his observation that IT initially provided a competitive advantage. But after mass adoption, all industries had harnessed its power, and IT became as essential to modern business as electricity had been to the industrial revolution. It had transcended its own uniqueness and become essential. If game technology is as successful, it will lose its niche status to become an essential part of running an effective and profitable business.

Chapter 14

A HISTORY OF GAMING IN MILITARY TRAINING

The military has been using games for training, tactics analysis, mission preparation, and systems analysis for centuries. Each generation has had to wrestle with the personal and public image of a game being used for something as serious as planning warfare in which people's lives are at stake. During the opening years of the 21st century the industry faces a renewed version of this question with the widespread use of computer games taken directly from the entertainment industry. For example, the AMERICAS ARMY project is based on UNREAL TOURNAMENT from Blizzard Entertainment and DARPA's DARWARS AMBUSH product is derived from OPERATION FLASHPOINT by Bohemia Interactive Games. Other agencies have converted DOOM and HALF-LIFE, or licensed products like GAMBRYO or the HAVOK physics engine to create military training systems. The questions, perceptions, and compromises that these projects face today are not new; they have been experienced by previous generations of engineers who attempted to leverage technologies from other domains. The mathematical models, paper board wargames, and miniatures that were adopted in previous centuries were met with the same type of apprehension that computer games face today. This article paints a broad picture of the history of the use of games by the military for training, the perceptual issues around that use, and the progress that has been made over many centuries.

MILITARY GAMING AGES

Stone Age

Simulation and gaming as tools of warfare has a very long history. Weiner (1959) believed that ancient Oriental generals may have planned their battles using icons on a map or scribbles in the sand. As far back as the Roman Empire, military leaders used sand tables with abstract icons to represent soldiers and units in battle. These allowed the leaders to visualize and manipulate a small physical copy of the battlefield. It allowed them to see information in geographic perspective and enabled multiple players to pit their own ideas against one another. Though the visual representation provided the initial value of the practice, the map or playing board upon which multiple options could be compared proved to be even more powerful. These tools allowed leaders and their staff members to compete against each other or against historical records in an attempt to determine which ideas would be the most effective (Perla, 1990).

Sand tables begat miniature gaming as both a military tool and a form of entertainment. Fred Jane, pioneer of the series of reference books on military weapons , created the "Naval Wargame" in 1903 as a military tool and shortly thereafter H.G. Wells, the famous author of *War of the Worlds*, published the book *Little Wars* in 1913 in which he described the use of miniature tokens and terrain boards for both military training and entertainment. From these roots sprung a century of the use of miniatures in planning military operations.

Paper Age

Strategy board games made of wood or paper emerged in Asia, the Middle East, and Europe. In Japan there are ancient references to the game of WEI HAI around 3000 BC, a term that meant "encirclement." This game is thought to be the predecessor of the modern game of GO (circa 2300BC). Both of these used abstract tokens that the player manipulated to gain a territorial advantage over an opponent. CHATURANGA emerged in India in 500BC with a gridded board and pieces that represent the leaders and warriors on a battlefield. It allowed two-player and four-player games and is the clear predecessor to modern CHESS. It used pieces which explicitly represented the military equipment of the time – the chariot, cavalry, elephant, soldier, and Rajah. In its original form a die was used to select the piece to be moved. However, Indian legal restraints on gambling and games of chance were applied to CHATURANGA, forcing its adherents to eliminate the die and allowing the players to develop their own strategies for selecting and moving pieces. By 500AD, this game had moved through the Middle East and Europe, being modified into the game of CHESS that we recognize today. Along the way many cultures cast it as the ultimate test of strategic thinking in a military context. The identification of specific pieces, the movement patterns assigned to each, the size of the playing board, and more advanced rules required centuries of experimentation to arrive at their current balance and to create a game that could challenge players for a lifetime and countries for centuries (Shenk, 2007).

Though GO, CHATURANGA, and CHESS may still be considered games of strategy, they are no longer accepted as training tools for warfare. The continuous evolution of gaming and training had led to more military-specific tools. In 1664 Christopher Weikmann created KOENIGSPIEL which was one of the earliest board games meant primarily for developing and communicating strategies of warfare. He was followed by C.L. Helwig with WAR CHESS in 1780 and Baron von Reisswitz with KREIGSSPIEL in 1811. All of these devices were tools to improve military thinking and to enable military training. During the late 19th century, the United States Naval War College used wargames to plan U.S. defenses against a British invasion of New York harbor. The Germans used wargames to plan the invasion of Poland at the beginning of World War II and the Japanese used this tool to plan the attacks on Pearl Harbor (Perla, 1990).

By the 1950's, two independent inventors introduced the paper board wargame with cardboard military markers and a table of combat results for calculating attrition and movement. The RAND Corporation created a system to present theater-level warfare in a form that would allow more mathematically precise and reproducible combat than that found on the sand tables and the board games of earlier centuries. They were charged with integrating the effects of the newly created nuclear arsenal, as well as developing strategies for defending against nuclear attack. Without direct experience with these weapons, wargaming was one of the few ways to explore their potential on the battlefield and in international relations (Ghamari-Tabrizi, 1995). At the same time, Charles Roberts was awaiting his commission in the Army and created a tool with which he could hone his own tactical skills. The result was a board game that he named TACTICS and which used many of the same techniques created

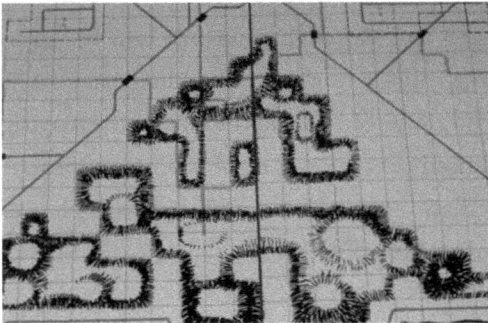

Figure 14.1. Charles Roberts' TACTICS II board game (circa 1958)

by RAND. The results of both efforts were the formalization of the playing board with a gridded overlay to manage movement and engagements; the use of a Combat Results Table to formalize the results of the battle; the incorporation of terrain types that influence combat activities; a turn-based play mechanism; and the use of dice to add random outcomes to the battle (Figure 14.1).

If these board games had come only from RAND, they may have been considered just a tool for military planning and training. However, Roberts used his creation to create the commercial entertainment company Avalon Hill in 1958 (today part of the Hasbro Company). He popularized wargaming as a hobby and a form of entertainment for those interested in trying their hand as military leaders. These games attracted a significant following of people who were well educated and may have had military experience. They found in this genre the opportunity to express their knowledge and to build small businesses based on their own creations. Hence was reborn the dichotomy between games as serious military tools and games as a form of entertainment. The concern of the appropriateness of playing games for serious purposes has been part of the education of military leaders ever since their popularization in the 1960's (Perla, 1990).

Ghamari-Tabrizi (1995) describes a situation during the Korean War when these board wargaming tools were relatively new tools for teaching strategy and tactics in the U.S military. The military colleges were using them to teach officers the craft of warfare. However, the image of using games as tools to educate military leaders was considered to be something the public was best not aware of. Therefore, the practice was kept secret. However, Milton Caniff, the artist and author of the extremely popular "Steve Canyon" newspaper comic strip had very tight relationships within the Air Force. Steve Canyon was a fictional America hero and a valuable recruiting tool for the military. Caniff learned of the use of wargames in these colleges and did a long series of comic strips in which the hero used them to plan his missions. Thus, the use of games for military training was exposed to the general public in the "funny papers" (Caniff, 1992).

Mathematical Age

The military wargames from RAND and Avalon Hill incorporated more arithmetic and stochastic algorithms than their predecessors. Players for entertainment sought a game that was easy to use, but military thinkers needed something that was more precise, even if it was somewhat cumbersome to play. This latter camp adopted computing devices to aid in their calculations. Calculators could be used to generate the results of specific actions and these results could be saved in table form for reference when the situation was encountered again during the play of a game. As computing calculators became more accessible they could be run in real-time to calculate specific combat results during play. This brought more detailed mathematics and logic to games and required no changes to the form of the game itself. Games continued to be played with paper boards, paper pieces, physical die, and miniatures.

During this period, computers could be viewed as an advanced form of calculator that was used by the more sophisticated military organizations. However, the cost and rare access to computers widened the gap between the "professional" and the "hobby" users of these games. This split encouraged hobby players to create affordable and easy to use manual mechanisms that could improve the richness and realism of their games without resorting to computers.

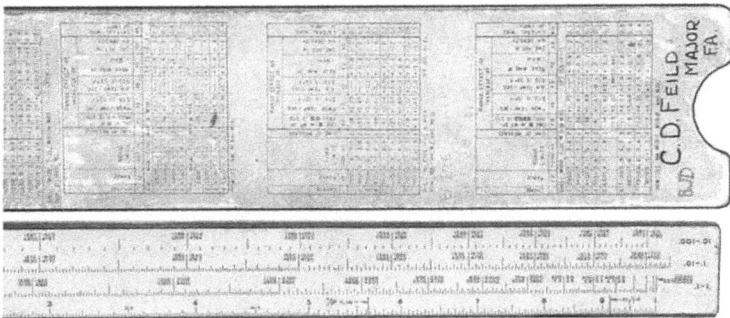

Figure 14.2. World War II firing table in the form of a slide rule.
Source: http://www.rekeninstrumenten.nl/

In 1948, the Army Operations Research Office at Johns Hopkins University created the AIR DEFENSE SIMULATION and in 1953 the first of a series of models called CARMONETTE. These games lost much of the playability of the board game, but significantly improved the mathematical precision and reproducibility of the results. These were some of the first truly computerized wargames (Davis, 1995).

Computer Age

Eventually, computers became powerful enough and employed display devices that would allow the entire wargame to be converted into a digital program. This eliminated much of the manual work of moving pieces, rolling die, looking up results in a CRT, and calculating final outcomes. The players could focus on the tactical movements and leave the complexity of manipulation to the computer. This also made it feasible to expand the size of games. The breadth of geography, number of icons, and complexity of algorithm no longer had to be limited by the necessity for human manipulation. The game could be expanded to the capabilities of the computer, which was many orders of magnitude faster than a human player with pencil, paper, and calculator.

Initially, digital wargames were a direct conversion of existing paper wargames. It took some time for the designers and programmers to discover the potential of the new computing machines and to develop more complex logic and algorithms for the computer. They could incorporate mathematic and logical algorithms that were far beyond what could be managed with a human-driven paper game. It also became practical to distribute the game across multiple rooms and to present custom views of the battle for each player (Allen, 1987). Prior to the computer, games with multiple boards and varying views were possible, but required the use of referees who could move between rooms to manually build up the unique views. These methods significantly slowed the play of the game. Distributed computer games moved this job onto the computers and networks, achieving almost instantaneous synchronization of multiple views of the battle. Allen (1987) describes the emergence of early games like the MCCLINTIC THEATER MODEL at the Army War College and the NAVAL WARGAMING SYSTEM at the Naval War College. The roots of these tools can be seen as early as the 1960s. Both games incorporated the most modern computers, networks, and display devices. They were followed by a long series of computer-based games developed for both training and analysis by the military services, governmental laboratories, and think tanks. IDAHEX was created by Paul Olsen as an analytical model that also allowed human player interaction. Paul Davis' RSAS system provided an environment for analytical combat outcome estimation.

These simulations improved the mathematical models of warfare, but they also began to bring in attractive graphics. At the time, the importance of display devices was not completely appreciated. This was one of the first steps toward bringing the military and hobby players of these games back together. In a previous generation, both communities had shared common map boards and pieces. In the convergence that lay ahead they would both be building systems on personal computers and the uncomfortable partnership between the two communities would be revived.

Personal Gaming Age

Today, personal computers and graphics cards have created an affordable platform for supporting games for entertainment, military training, medical education, and global communication. Though the entertainment and military communities had diverged through the 1970's and 1980's they became reacquainted toward the end of the 1990's. Military simulations like Simulator Networking (SIMNET) and Modular Semi-automated Forces (MODSAF) were uniquely military and proved quite valuable in preparing Army officers for actual maneuver warfare in the Desert Storm war (Miller & Thorpe, 1995).

These were leveraging the graphic tools emerging from academic research laboratories and commercialized by companies like Silicon Graphics and Evans & Sutherland. On the entertainment side, games like STEEL PANTHERS and CLOSE COMBAT carried a very strong military theme, while SIMEARTH and CASTLE WOLFENSTEIN went off in a uniquely entertainment direction.

The growing power of the personal computer allowed some of the military simulations to migrate from large institutions to individual hobbyists. Traditional military training games like JANUS and parts of SIMNET were recreated in a gaming form and sold for entertainment. The SPEARHEAD game was jointly created by an entertainment company and a defense company which was comprised of some of the original developers of SIMNET. At the same time, entertainment games and the technologies behind them, seeped into the military domain. Blizzard's UNREAL TOURNAMENT became the foundation for the sensationally popular AMERICAS ARMY game for recruiting (Zyda, 2003). FULL SPECTRUM WARRIOR was created for the X-box console as both a military training system and an entertainment title. OPERATION FLASHPOINT became the basis for the DARWARS AMBUSH training game (Chatham, 2007).

Throughout these decades long evolutions people have been concerns about turning entertainment properties into military applications, often referred to as "serious games" (Zyda, 2005). There is the concern that games may be adopted because they are visually attractive rather than accurate representations of battlefield activities. These same concerns have arisen in all industries that adopt technologies that originate from other industries. For example, the medical education field has done numerous studies to determine whether simulations of all types can provide better training than traditional hands-on methods. In most cases they have learned that these new training tools are essential for teaching the complex skills that are necessary for the latest forms of surgery and the use of advanced equipment (Lane, 2001).

Game Technologies

Entertainment games encapsulate some of the most advanced technologies in computer science. They present compelling and powerful tools that are being adopted by many industries, including military simulation. This section explores the power of these technologies in order to better understand what they offer for military training. Figure 14.3 shows six technologies that are core to the runtime experience of a game-based training system.

1. The *3D engine* creates the beautiful and precise visualizations that are used to stimulate players.
2. The *Graphical User Interface* includes the menu system and interactive patterns that allow a gamer to immediately begin using a game without reading the manual.
3. *Physical Models* create a world in which the effects of movement, engagements, interactions, and sensors are accurate portrayals of the real world.
4. *Artificial Intelligence* provides the brains that are necessary to create in-game opponents that are smart enough to challenge human players. This also creates an adaptive experience that can adjust the game as it runs to insure that the player works through a specific problem of interest.
5. *Networking* allows the virtual environment to be extended to multiple players around the globe. Game companies have created the most flexible and efficient methods for transferring data over a worldwide network.
6. *Persistent Worlds* maintain a virtual world that grows, changes and remains active over many days, weeks, or years regardless of whether an individual player remains engaged with that world.

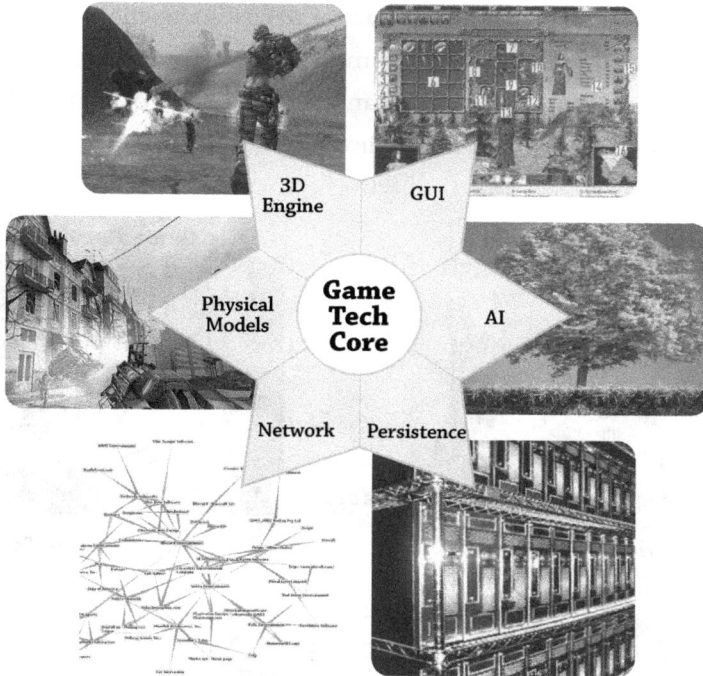

Figure 14.3. Six core technologies in gaming systems

But the technologies of in-game play are just one part of what is driving these systems into serious applications. Most serious domains call for a great deal of pre-game data creation and manipulation, as well as post-game data collection and analysis (Figure 14.4). *World Building Tools* allow the training audience to customize their own scenarios to meet very specific needs. With these a player can create new pieces of terrain, buildings, vehicles, and complex layouts of these to form a game level. *Behavior Scripting Tools* enable the creation of specific actions and motivations for the enemy avatars that are controlled by the computer. Prior to the availability of these types of tools the original designers of the game had to determine how it would be used and create scenarios that could teach those lessons. It was just too complicated for players to modify a world themselves (Chatham, 2007). However, the latest generation of games includes tools that allow any user to customize the world and create scenarios that will unfold during a game. Post-game tools collect data on the performance of each player or team and allow a review of that performance with instructor comments on how to improve.

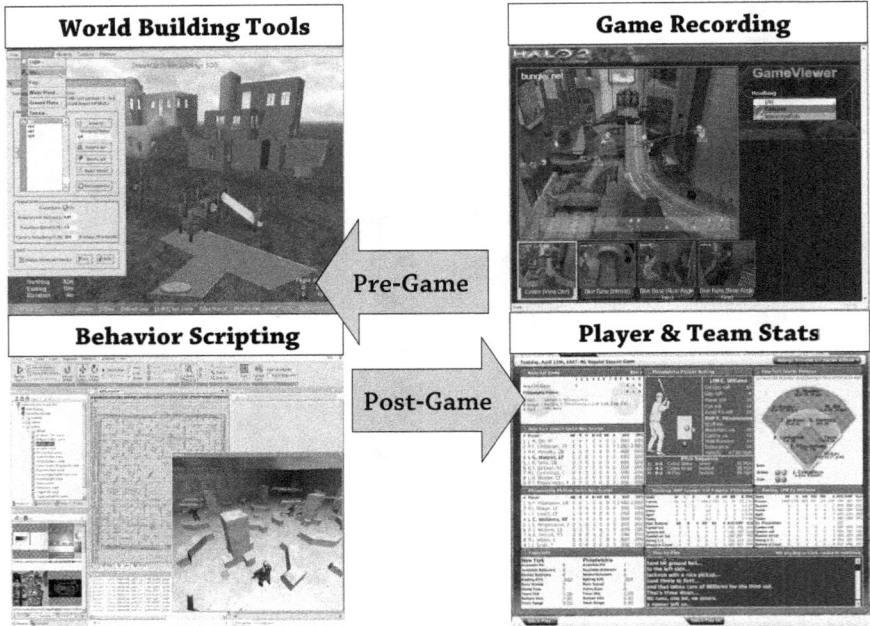

Figure 14.4. Valuable technologies for pre- and post-Game operations

As games have become more sophisticated and as the military has come to understand them better, we have been able to identify better means of leveraging these technologies for serious purposes.

FUTURE TECHNOLOGIES

Games of warfare have evolved into multi-processor systems that typically use a client-side application to deliver the user interface and to handle computation that is dedicated to a single player. They also have a server-side application that can handle more globally applicable information and that ties multiple players into a common environment. As the wargame becomes more complex it usually requires additional power on the server side. As we seek to make the system more intuitive to the players, it requires more computation on the client-side. These demands point to new computer and software technologies that will be important in the future.

On the server side, the growth in computational needs is similar to that found in large IT server farms that drive Internet applications like Google Search, Maps, and Earth; but also those that drive massively multiplayer online computer games like WORLD OF WARCRAFT, SECOND LIFE, and EVE ONLINE. The former are well structured for customers who submit independent tasks that do not have to be coordinated in real-time with tasks from other customers, such as word searches and requests for map directions. The latter support a series of small virtual worlds in which thirty or more players may share the same virtual space and need a server-side computer than can coordinate their actions in real-time. The "massively multiplayer" part of these games is created by running thousands of these small groups simultaneously and independently. In spite of the large number of players that are reported, relatively small numbers of these work together in a shared space simultaneously. If a military application is going to support hundreds of players who interact with each other, then something more powerful will be required—perhaps, something like an interactive version of a supercomputer (a.k.a. high performance computer). These machines have typically been used to handle computationally intensive jobs that do not require interactivity with the customer. The large number of processors, powerful chips, and large memory spaces could also be used to support large interactive exercises if the connections to the players were maintained throughout the job, rather than being terminated as is typical of "batch jobs".

As computers have become ubiquitous across the soldiers, sailors, airmen, and marines that make up the military services, it has become possible to envision an environment in which these customers or players can use their own computers as the clients in a global, server-based training system. Millions of potential trainees or players could connect to servers to run collaborative training on their own schedules. The servers would deliver specific content to each player according to his or her role and the skills that they are trying to learn. It would be impossible to manually deliver specific content to each of the millions of servicemen and women around the world. Instead, the servers would have to supply the necessary content and connections on demand. The means of establishing such a system are currently beyond the capabilities of the military networks, servers, and clients that are available. But the shadows of such a future can be seen in the demands for custom training, the shift to constant operations, and the types of innovations that are emerging in the commercial spaces for games and internet-based interactions.

ECONOMICS OF THE CUTTING EDGE

At the beginning of the 21st century there is concern that commercial and entertainment technologies appear to be outstripping those in military systems. It appears that we have reached a global economic inflection point in which the populations of many countries have significant disposable income and disposable personal time. These people choose to spend this surplus on personal computing equipment and games for entertainment, creating a gaming industry that earns over $18 billion annually. This industry is infused with huge amounts of money each year and generates an environment of extremely aggressive competition. Existing and newly formed companies introduce more advanced hardware and software products every year. This revenue and the competitive environment have seized the position of technical leadership from the military and government organizations that held it for decades. There was a time when the most advanced computing devices could be found in military weapons and vehicles for space exploration. Today, the revenue behind commercial computing hardware and software is much larger than that available to the government. Computer games and similar entertainment products are just one part of this larger technology shift. The military has become accustomed to using commercial off the shelf (COTS) computer hardware. It must look forward to using COTS software for applications like training and data analysis. Three dimensional image generators used to be cabinet sized machines. Today they are small cards that fit into a personal computer. Maps used to be created on paper by professional cartographers. Today they are embedded in web pages and can be customized by anyone surfing the web (Figure 14.5).

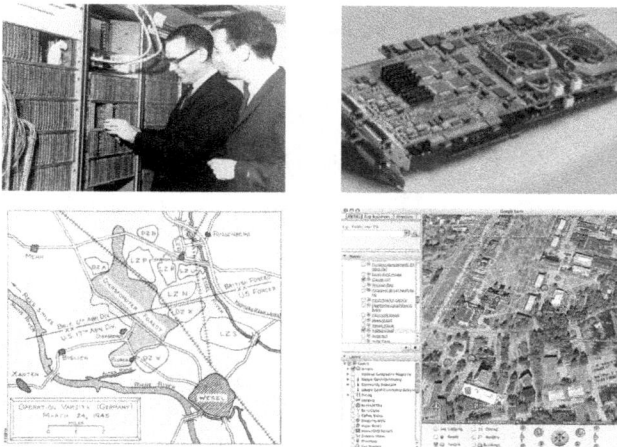

Figure 14.5. Commercial products can completely displace military-specific hardware and software

CONCLUSIONS

The term "game" refers to an activity pursued for entertainment and enjoyment. Tt also denotes a competition between multiple players. This competition reaches its ultimate form in military operations where life, liberty, and global security are at stake. These games are not played for entertainment, but they incorporate the competition and strategies that emerged from their entertainment roots. Since games can be used to think clearly about military operations, they have been and always will be a part of this very serious activity.

Simulations, games, and virtual worlds are also spreading beyond their roots in training and analysis. They are becoming part of the global communication infrastructure. They are tools that can be used to think about all kinds of problems and to orient dispersed audiences toward a shared problem in three dimensions.

Chapter 15

COMPETITIVE IMPACT OF GAME
TECHNOLOGIES ON FIVE INDUSTRY

In his Nobel Prize winning work, Robert Solow demonstrated that the advancement of technology has been responsible for 50% of the economic growth of nations in the 20th century. The application of these new technologies created new businesses and transformed established institutions, allowing them to be significantly more productive than they had been prior to applying the technology (Solow, 1957).

Computers have been one of the most significant technologies introduced into all institutions. Computers are computational, information handling, and telecommunications devices, all of which have brought significant innovations to the 20th century. Newspapers had been around for centuries as a primary means of sharing information with a large audience. But this all changed around 1920 when our understanding of the properties of radio waves allowed us to create radio broadcasting stations and news was suddenly turned into an electronic form that could be transmitted almost immediately into people's homes. Similarly, the performance theater lost its place as the primary delivery vehicle for acting when film and radio joined forces to enable broadcast television. Digital signal processing and miniaturized data storage digitized the music industry and enabled Internet downloadable music and the MP3 devices that play it.

Detailed computer graphics and animation had been relegated to academic laboratories and major military programs until companies like Evans & Sutherland, Silicon Graphics, 3dFX, and Nvidia introduced the computer graphics processors. Their success in the market and improvements in computer chip manufacturing led to the creation of a number of affordable graphic devices for all forms of computers. Today anyone can install a major graphic processor in their computer for less than $100. This has enabled the visualization, animation, texturing, shadowing, and lighting of a three dimensional world for use in any application with a need. The leading users of this capability have not been scientific researchers, but instead it has created explosive growth in the computer game industry. In a single decade it has risen from relative obscurity to a size that rivals that of the motion picture industry. Zyda (2006) states that the worldwide game industry was $33.5 billion in 2005 and may grow to $58.4 billion by 2007. As the industry has grown, the tools and technologies have struggled to overcome their stigma as a game or toy for entertainment. We are witnessing the early stages of adoption of graphics and game technologies into a number of mainstream industries. As early as 1970 Charles Abt labeled this trend "serious games", the title by which it is known today (Abt, 1970).

The early stages of adoption of such a powerful technology are attracting the attention of business, academic, and government leaders. It has also become a popular focus in the media in the last few years. It is not uncommon to see a new story on the use of game technologies for serious purposes in leading business publications every week. This trend and the changes it promises appear to be poised to significantly impact the economies of a number of developed and developing countries. Just as India has become a serious player in providing information services, some country or cluster of countries is going to become a provider of serious game technologies. Currently this expertise is emerging in the United States, but once the business model has matured and a solid market is identified, other countries will focus their intellectual energies on this market as well. Several Eastern European countries, former Soviet states, possess significant intellectual talent and low operating costs that have allowed them to create some very successful game titles and to build a reputation as a potential development cluster for game technologies.

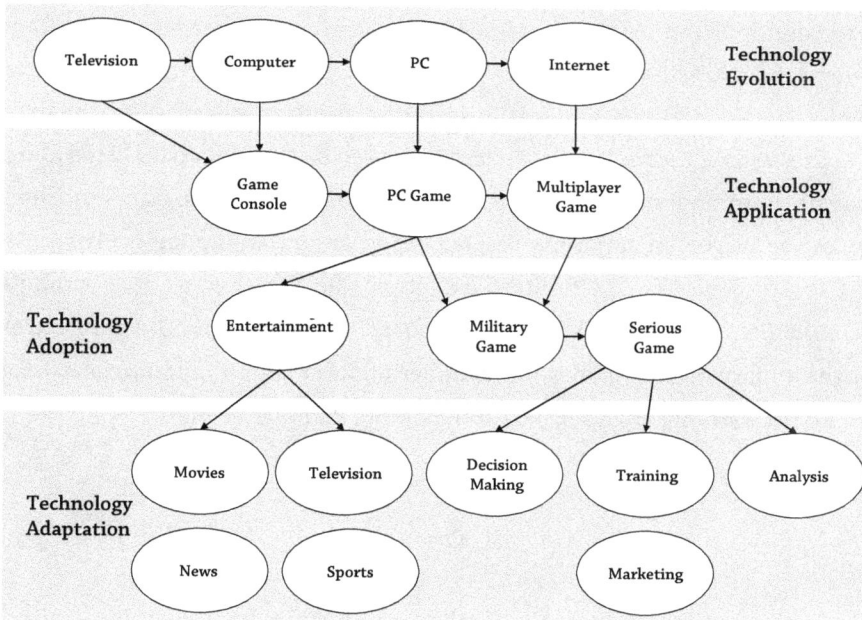

Figure 15.1. The emergence and adoption of necessary technologies to create and drive the global explosion in game products and services.

In this paper we will explore the application of game technologies to other industries, focusing specifically in five categories—entertainment, training, scientific analysis, decision making, and marketing. We will begin by examining the business literature for several broad themes that are related to the transformation of industries. Understanding the forces of economic change is important if we are to understand the potential that game technologies have to initiate a new wave of change.

Figure 15.1 illustrates a path of technical innovation and transformation that has created the commercial and serious games industries. The evolution of technology has brought us the television, main frame computer, personal computer, and Internet. As they became available to large audiences, each of these have been applied to electronic entertainment, first in the form of game consoles, then as PC games, and finally as multiplayer games leveraging the Internet. These game technologies are being adopted by the industries listed above. The entertainment industry includes much more than games. Game technologies are appearing in major movies, television programming, news broadcasts, and sporting events. The military has always be a user of leading computer technologies, but as they adopted computer games they laid the groundwork for the applications of games to a number of other serious applications—specifically in training, analysis, decision making, and marketing.

RELATED LITERATURE

Creative Destruction

In 1942 Charles Shumpeter wrote *Capitalism, Socialism and Democracy* in which he discussed the long-term viability of a capitalistic economy (Shumpeter, 1942). In that book he also described the impact that change has on business and the economy.

> *"We must now recognize the further fact that restrictive practices ... as far as they are effective, acquire a new significance in the perennial gale of creative destruction, a significance which they would not have in a stationary state or in a state of slow and balanced growth."*

This description and one phrase in particular, have become famous among modern economists and business leaders. Shumpeter cast a positive light on "creative destruction", the impact that change, technology, and knowledge have on the current state of the economy. An economy that remains static or grows slowly is not in the clutches of the forces of growth. When significant growth is occurring it will lead to the destruction of existing structures and the devaluation of skills that were previously essential. This is a positive force that creates more opportunities than it destroys. However, it can also be very painful because most people and businesses cannot clearly see where the new value is being created and must often suffer the loss of income and jobs before they can enjoy the fruits of renewal and growth.

Shumpeter's ideas were reinforced by the writings of David Wells, who insisted that:

> *"Society has practically abandoned—and from the very necessity of the case has got to abandon, unless it proposes to war against prog-*

ress and civilization—the prohibition of industrial concentrations and combinations.

"It seems to be in the nature or natural law that no advanced stage of civilization can be attained, except at the expense of destroying in a greater or less degree the value of the instrumentalities by which all previous attainments have been affected."

(Perelman 1995)

Wells recognized the necessity of abandoning past practices in order to achieve new and more powerful methods of building an economy. Wells describes the necessity of "industrial concentration", what we would call mergers and acquisitions today. He realized that the overcapacity that is generated within any successful industry is damaging to that industry and to the economy as a whole. Overcapacity robs the economy of the productivity that could be realized by applying those human, financial, and industrial assets to a new and different effort (McDaniels, 2005). When 50 people are employed in a company that really only requires the labor of 25 to do the work, the company and the economy suffer for it. The company experiences higher costs and lower profits. This leads to the customer paying higher prices for their products. If the 25 surplus workers were removed from the payroll, then the company would be able to make its products at lower costs and sell them in the market at lower prices. In addition, those 25 people would become valuable assets in some other industry that needs their services. In this way the general economy benefits from lower prices and higher productivity from its population. When viewed on a national scale, this helps us understand why countries that encourage the retention of labor even when there is no need for them in an enterprise, cannot experience the growth levels that we see in more flexible and dynamic economies. The assets in static or regulated economies are tied up performing lower-value operations and cannot be freed to stimulate growth.

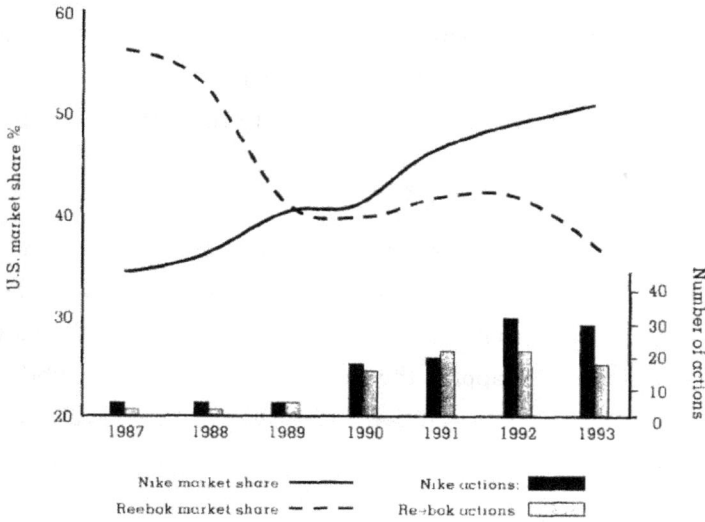

Figure 15.2. Action aggressiveness between Nike and Reebok
Source: Smith 2001

Smith (2001) studied the impacts that changes have on the competitive positions of companies. He found that, *"A creative action carried out by challengers disrupts the competitive status quo and forces industry leaders to respond and also to be creative."* Further study indicated that, *"The more new competitive actions a challenger takes, the greater is the competitive uncertainty on the part of industry leaders, which will cause a delay in their reactions."* This indicates that maintaining the status quo in a business can lead to the loss of a leadership position to a competitor who is willing to race from one change to another. Those who change slowly will find themselves reacting to the change leader and may find that they lose their market leadership to the change leader. Smith demonstrates this through studies of a number of companies in different industries. Figure 15.2 compares the market shift between Reebok and Nike in the sports shoe business. In the late 1980's Reebok held a significantly larger market share than Nike. However, Nike began to introduce innovations to its products, partnering, and marketing. It made more changes per year than Reebok. During this time there was a corresponding shift in market share away from Reebok and toward Nike until in 1993 the two had nearly reversed their positions from 1987. This illustrates Smith's point about the significant impact that business changes can have on the balance of leadership in an industry.

Reinganum (1985) also points out that challengers always invest more in innovation and changes than do incumbents. This means that monopolies are often short lived because a challenger will soon innovate and spend their way into a leadership position. She applies the Nobel winning "Nash Equilibrium" from Game Theory to demonstrate that these innovators are more likely to win competitions among "n" players in non-zero sum games like business competition.

Popular authors in the business press often turn to Shumpeter's ideas to both justify and explain the temporary negative impacts that come from technological change. These impacts are both financial/economic and personal/ emotional, which leads to the next set of related literature surveyed for this paper.

Change Management

Change management authors focus on the disruption, confusion, fear, and insecurity that many people face when significant changes occur in their business. Creative destruction may be an improvement in the economy, but it is not an improvement in the mental and emotional state of the people who are being disrupted. Managing change often boils down to managing the impact that this change has on people.

Jones (2004) emphasizes that change management has a very human side and that effectively addressing this requires that leaders from the top of the organization participate in the process of educating people about the change and their place in the structure of things. The change process must include everyone in the organization. It must become something that they are participating in, not just something that is happening to them, or something that is being done to them.

Change management is directly relevant to the application of game technologies that are going to displace tasks, tools, and workers that are established in a number of industries. A company is either going to have to restructure, retrain, or restaff itself in order to remain competitive. Like all major changes, these are going to carry significant emotional impacts to employees. This is an issue that must be handled carefully. An organization in the midst of significant change can find that it's best and brightest people leave when they were the very people upon which the company was planning on building the new organization. Averting this requires frequent and voluble communication with the people, both those who will stay and those who will leave.

A major part of communication is creating a shared vision of the future of the company, the industry, and the role that each individual can play in this. The vision of a new type of future cannot be a private picture to the leaders. Instead, it has to become a shared picture that everyone feels some ownership for and sees a role for themselves in (Jones, 2004 and Conger et al 1999).

Conger et al (1999) reports the research of Mohrman and Tensaki in which they identified seven positive impacts of change on people in an organization. First among these was *"an opportunity to learn new things"*. Another was *"an opportunity for career growth"*. These two are probably dominant in the injection of game technologies into an industry.

Conger also provides eight discrete steps that should be included in transforming an organization (Conger et al, 1999, p.99):

- **Establish a sense of urgency.** Leaders and employees in the organization need to appreciate the threat that is posed by new technologies.
- **Form a powerful guiding coalition.** Create a group that actually has the power to address the issue and make real organizational changes.
- **Create a vision.** Identify what the future looks like for the company.
- **Empower others to act on the vision.** Make it possible for people at all levels to move themselves and their sub-organizations toward the new vision.
- **Plan for and create short-term wins.** Capture successes, learn from them, and advertise them. Make it clear that progress is being made.
- **Consolidate improvements and produce more change.** Restructure if necessary to leverage improvements. Hire people who can contribute to the new organization.
- **Institutionalize new approaches.** Change the reward structure to encourage new behaviors. Create a new mythology around the repurposed organization and its employees.

Tushman and O'Reilly (1997) identify four major areas in which change is applied—Critical Tasks, People, Culture, and Formal Organization. Many leaders are quick to change the formal organization and to begin identifying the new set of critical tasks. These are the mechanistic side of change. In recent years, managers and leaders have also developed a new appreciation for addressing the needs of the people. Tushman and O'Reilly argue that changing the culture is often overlooked and contains the potential for making more significant contributions than the other areas. Culture defines what is meaningful and valuable within the organization. As long as the perceived value remains unchanged, it will be difficult to get people and groups to adjust their behavior to match the new goals of the organization. Tushman and O'Reilly point to symbolic actions, rewards, and recognitions as key to realigning the culture of an organization. All of these communicate to the employees what the company values in behavior, attitude, and action. If these do not change, then the employees will not change their behavior regardless of the communicated signals.

Technology Innovation and Disruption

"Technology disruption" has become the "creative destruction" of the early 21st century. Ever since Clayton Christensen put his finger on the trend in his 1992 dissertation it has been a favorite topic of academics, consultants, and writers. He refueled a discussion that has always been part of the management literature and practice.

Christensen (1997), Utterback (1996), Leifer (2000), Subramanian (2004), Thomke (2003), Hardagon (2003), and von Hippel (2006) are just a few academics who have been studying the impact that technology and innovation are having on the industrial and information ages. Each has presented the picture from a different perspective and communicates the message that companies, industries, and countries can no longer maintain a static position in the global economy. We are at a point in history when technology is leveling the playing field for thousands of companies and hundreds of countries. This is allowing new products and services to rise from obscurity to achieve industry or even world dominance in spite of the efforts of established leaders to hold onto those positions.

Currently, it is doubtful that stability exists in any industry that has attractive revenues and profits. China, Asia, India, and Eastern Europe are all being equipped with the technology, finances, information, and education necessary to compete for business that has traditionally been dominated by the United States, Canada, the United Kingdom, and Western Europe. Industry leaders have been toppled and others will continue to fall as the balance of wealth shifts to match the new balance of industrial productivity.

We have emerged from a time in which established companies offered incremental innovations to their products every year or two. Customers had learned to live with products as they were because no major changes would be appearing for years. But today, we see that much more radical innovations are happening in a number of industries and they are happening at a rate that is unprecedented in previous years. In the last decade, the traditional wired telephone service has been significantly challenged by mobile cellular networks. Just as the cellular infrastructure is becoming reliable, we see that computers are beginning to replace expensive cellular services with nearly free telephone service through wired and wireless connections. On the horizon is a hybrid phone that can nimbly jump across all available communications mediums, moving from a wired connection, to a wireless computer network, to a cellular network, and back to wired all while maintaining a single continuous conversation from the perspective of the customer (Ashley, 2006).

IBM's study of global CEO's uncovered some interesting opinions about the sources of new innovations in products (IBM, 2006). The executives interviewed did not believe that significant new products would be initiated by their own research and development departments, or even from their corporate leadership. Instead, they felt that innovative products would be created and proposed by customers. In line with the work of Eric von Hippel (2005), they believe that a small group of radical or leading edge customers will transform products to meet very specialized needs. And they believe that these specialized needs portend the needs and desires of the central mass of their customers in the future. Therefore, the R&D branch of major companies has migrated from the laboratory to the extreme edge of the customer base.

Clayton Christensen (1992, 1997, and 2003) demonstrated that young start-up companies can potentially overturn the leadership of much more established companies. Because established companies often discount the value of new technologies that do not immediately exceed the ability of existing technologies to meet the needs of existing customers, they often dismiss opportunities that have the seeds to grow into more powerful products and pull a significant customer base toward them. His studies of the evolution of computer hard drives showed that the same phenomena have occurred over and over, even when the current leader arrived at their position by leveraging new technology to overturn a previous incumbent.

It appears that game technologies may have the potential to do this same thing to a number of established industries. Though perceived as having inferior capabilities in many serious applications, their advantages in visualization, speed of development, price to deploy, level of expertise to use, and acceptance by younger professionals may catapult them into a major force, in spite of their initial weaknesses.

James Utterback (1996) presented a model with three different phases of innovation that occur as an industry or product family evolves. Initially, innovation, creativity, and experimentation are focused on the form and function of the product. Given a new technology like handwriting recognition, digital photography, or cellular communications, it is not immediately obvious how to package this into a product that will attract consumers. In fact, since customers are completely uninformed about new applications, it is impossible to extract this information from them. Instead, a number of companies create different forms of products, gradually coming to understand the needs and desires of customers. At some point a dominant design emerges that appeals to customers, meets their needs, and stands as a symbol for the entire product area. Once this occurs, innovation shifts from product form to product creation and delivery. The competitive space shifts to manufacturing speed, cost, and ergonomic design. There is a significant drop in the basic variations in the product and an increase in new package designs and product accessories.

When near-optimum product design and manufacturing is achieved, the product moves into a commodity status. Innovation now focuses entirely on costs, advertising, and convenience of access.

Computer games for entertainment went through these phases and settled on a few core categories of products—real-time strategy, first person shooter, simulation, puzzle, learning, and a few others. Once this occurred, the focus was on the ability to deliver product modifications on a regular basis. Games are on the threshold of becoming commodities. High quality content will be available at many outlets for prices below twenty dollars. Fighting against this commoditization, the game industry continues to push forward with visual detail and the breadth of the virtual world. Significant investments in advertising are required to convince customers that the newer products are superior and justify game prices of $40 to $50.

Leifer et al (2000) point to the importance of radical large company innovation to remain ahead of ambitious start-ups. Their research is a reaction to the dire picture painted by Christensen in *The Innovator's Dilemma* (1997). Some of America's largest companies worked with the authors to identify innovation practices that could help them maintain their lead. The prescription for radical innovation emphasizes the reinvention of products, processes, facilities, and staff that are core to these companies. The era of incremental change to long-established products has ended. Customers are willing to switch from multi-generational brands like Tide, Colgate, and Kodak when they see something new, unique, and interesting. Established companies can no longer assume that these branded products are a permanent part of the national economy.

Von Hippel (2003 and 2005) explores the role that "lead users" play in changes to products. In the open source software and wind surfing communities, he discovered that users on the leading edge of the customer base experiment with products. They make changes to suit their "extreme" behaviors. Very often these modifications point to the most valuable innovations for the next generation of the main-stream product. Extreme users receive significant attention in the media and on the Internet. Their use of the product spreads through the community and creates demand for the modifications that they have made. Von Hipple suggests that companies need to connect to these lead users and treat them as an external research and product development team. In the game industry, lead users were those who played online games competitively. That niche has grown to a point that a single title now boasts a subscription base of over seven million players.

Thomke (2003) illustrates the financial importance of experimentation. In developing new products, making and identifying mistakes early is both more cost effective, and leads to faster product deployment. Thomke points to the essential role that experimentation plays in accomplishing this. Many formal product development processes move slowly and methodically, attempting to avoid mistakes. Unfortunately, this risk adverse behavior delays the creation and release of new products and does not allow sufficient interaction with the customer base. As a result, the risk-averse, non-failing products arrive in the market late and miss the sweet spot of customer desire. A more effective method is to use experiments to make mistakes early and often. This leads to faster product development and lower costs. This model fits well with the two year game development cycle in which new software techniques are incorporated into products almost immediately.

Hardagon (2003) describes the importance of people and companies that bridge two different industries. Dual-citizenship allows them to escape many of the limitations of both of the communities. When business is poor in one area, they have the flexibility to focus on the other. But this also allows them to break with traditions and stereotypes from a single industry. They are more likely to pull ideas from one domain into another, which can appear to be radical, creative, and completely spontaneous. Serious games are exactly such an application. These move technologies from the game industry to serious applications in the military, medical, architectural, and educational fields. Ideas that may seem routine in the game industry, like completing product development in less than two years, or reusing a common core of software from one product to another, often appear radical and ground-breaking in older industries.

Technology Leadership

Innovation, risk-taking, and new products do not emerge from the wishful thinking of company leaders. They require active leadership to encourage new ideas and to bring them to fruition. To achieve this, many companies have changed the focus of their research and development labs. In the 1950s these were seen as places where great leaps in technology and science occurred. As such, they were often located on remote campuses and staffed with brilliant researchers, creating an environment that mimicked that of a university. However, this model did not tie R&D directly to corporate performance. Too often it allowed researchers to pursue ideas that had no financial return. With the rapid growth of commercial technology and global competition, major companies could no longer afford to spend money on research without some promise of a commercial product. Therefore, these research labs were brought closer to the production facilities, both literally and organizationally. From a executive leadership perspective, this was expressed through the creation of a new position—the Chief Technology Officer. This role grew out of the established position of Director of Research, but extended that older title to include direct accountability for generating new products and services (Lewis and Lawrence, 1990).

The CTO joined the executive ranks and was charged with turning research or commercially available technologies into revenue-generating products. This was just one of many changes that began to move through organizations in the late 1980's and 1990's. CTO's like Nathan Mvyrold at Microsoft started "virtual worlds" research groups to investigate the degree to which three dimensional worlds could be incorporated into non-gaming products and to push the research edge of gaming forward. This research emerged several years later as two cornerstone products for the company, the Xbox gaming console in 2000 and the upcoming Vista operating system in 2007. Both of these products leveraged powerful research and express the new relationship between the research labs, products, and executive leaders.

As it has grown into an industry giant on the scale of IBM, Exxon, and Wal-Mart, Microsoft has struggled to maintain its ability to innovate new products and to remain a nimble competitor against the many established and start-up companies that are seeking to edge into their market space. This has led founder and Chairman, Bill Gates to step down from his executive role and take the title of Chief Architect. This move recognizes that the best minds and the most influential people are needed to seek out new products and new customer needs that transcend the traditional Microsoft mold. The Xbox and the new Zune music player are these types of products.

Effective technology leaders possess a number of complementary skills that enable them to play an effective role in guiding change in an organization (Smith, 2003).

- Technology. They understand the core technology that is driving company products and that is being introduced by competitors.
- Strategy. They have a focus on the strategic meaning of changes and look for ways to add technological support to corporate strategy.
- Business Growth. They share responsibility for the growth of the business, not just for the technology of the products and services.
- Interpersonal Skills. They lead, inspire, and control people effectively.

- Executive Relationships. They have strong working relationships with the entire executive team.

SPREAD OF GAME TECHNOLOGIES

One of the earliest published books on "serious games" came out in 1970 from Clark Abt, an MIT graduate who was focused on improving the educational system. In that book, Dr. Abt introduced the term "serious games" which has been adopted more recently in reference to computer-based games. Though Abt's games were all based on manual role playing or traditional board games, the concepts behind using entertainment techniques for a more serious purpose was identical to the transformation that is emerging from computer game technologies today. Abt describes "serious games" in this way:

"Reduced to its formal essence, a game is an activity among two or more independent decision-makers seeking to achieve their objectives in some limiting context. A more conventional definition would say that a game is a context with rules among adversaries trying to win objectives.

"We are concerned with serious games in the sense that these games have an explicit and carefully thought-out educational purpose and are not intended to be played primarily for amusement."

(Abt, 1970, p.9)

As computer games became more common as tools for serious industries, a number of authors attempted to create more descriptive definitions for the term. Zyda (2005, p.25-26) approached the challenge with a series of definitions leading up to one for serious games.

Game: *"a physical or mental contest, played according to specific rules, with the goal of amusing or rewarding the participant."*

Video Game: *"a mental contest, played with a computer according to certain rules for amusement, recreation, or winning a stake."*

Serious Game: *"a mental contest, played with a computer in accordance with specific rules that uses entertainment to further government or corporate training, education, health, public policy, and strategic communication objectives."*

Michael (2006, p.17) defines it this way:
"a serious game is a game in which education (in its various forms) is the primary goal, rather than entertainment"

Michael also provides a definition from Bernard Suits' book, *Grasshopper: Games, Life, and Utopia*:

"To play a game is to engage in activity directed towards bringing about a specific state of affairs, using only means permitted by rules, where the rules prohibit more efficient in favor of less efficient means, and where such rules are accepted just because they make possible such activity."

(Michael, 2006, p.18)

As early as 1970, Abt was able to identify a number of industries within which to apply games.

"Education, industrial and governmental training, planning, research, analysis, and evaluation are all rich fields for the use of serious games"

(Abt, 1970, p.10)

In a previous paper I compiled a list of industries that were known to be using games and game technologies to push their products and services forward (Table 15.1, from Smith 2006).

Table 15.1. Industries impacted by games and game technology.

Industry	Game Technology Impact
Military	Training soldiers and leaders in the tactics and strategies of war. Three dimensional modeling of equipment to illustrate or explore its capabilities.
Government	Ethics training for NASA. Project management training for the State of California.
Education	Augmenting classroom instruction in nearly every subject – English, math, physics, history, etc.
Emergency Management	Training emergency responders, firefighters, FEMA agents, and others to deal with disasters.
Architecture	Visually promoting major hotel, casino, and office spaces to potential clients.
City & Civil Planning	Lay out and experimentation with public services for a population of constituents.
Corporate Training	Orienting people to company products, facilities, and policies. Pilot and safety training.
Health Care	Educating patients on treatments, rehabilitation, and managing anxieties. The next generation of workout videos.
Politics	Presenting political issues and consequences of political decisions. Promoting candidates.
Religion	Interactive versions of sacred texts. Tools to teach religious history.
Movies & Television	Tools for creating animation and 3D worlds. Alternative form of storytelling known as "machinima".
Scientific Visualization & Analysis	Rapid display of objects under experimentation and physical forces acting on them. 3D display of data collected and analyzed.
Sports	Recreate live sporting events for review and for prediction of potential outcomes. Design and rehearse critical "one time" events like Olympic ceremonies. Fantasy sports leagues in 3D.
Exploration	Prepare missions for NASA Mars Lander. Recreate environments around deep sea probes.
Law	Illustrate crime scene activities for judge and jury. Analyze crime scene data.

Sources: Compiled from Michael & Chen, 2006; Bergeron, 2006;
Casti, 1997; Maier & Grobler, 2001

Mike Zyda lists seven root applications for serious games and a number of additional specializations off of the "training and simulation" root (Figure 15.3).

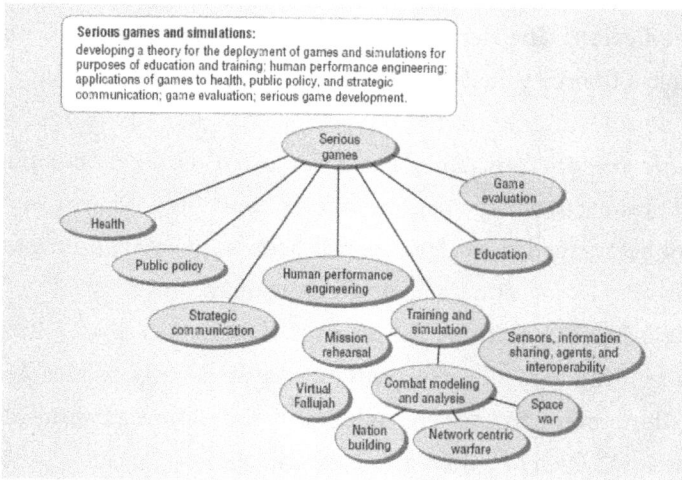

Figure 15.3. Zyda's taxonomy of serious game applications.
Source: Zyda 2005

In his 1968 work, Elmagharaby identified five applications of simulation techniques and technologies. At that time and from Elmagharagy's perspective in operations research, the term "simulation" largely referred to languages for discrete event simulation and mathematical techniques for optimization. However, in spite of that, these five areas continue to be very descriptive of the modern use of simulation and can be applied to the serious games business as well. Figure 15.4 illustrates Elmagharaby's uses for simulation and adds one that was perhaps not viable in 1968.

- **Aid to Thought.** They apply a computer's memory and ability compute on a massive scale to the human's very limited ability to do these same operations. As a result, humans are able to think about larger problems while maintaining accuracy and consistency.

- **Communication.** Simulations can assist in communicating ideas from one person who understands them to another person who is seeking understanding.
- **Prediction.** The accuracy and logical construction of simulations makes them a valuable tool in predicting the future state of a system or a problem.
- **Experimentation.** Simulations are often used as a computer-based laboratory in which experiments are conducted. Many problems cannot be pursued in the "real world" because of safety limitations or our inability to collect data in all circumstances.
- **Training.** Simulations are an engine for replicating a situation and presenting it to a human audience as a challenge to be solved. Many authors point to the repeatability of simulation as an ideal environment in which to learn with feedback.

The sixth application of simulation that is added here is as a tool for entertainment—a.k.a. gaming. As technologies become affordable, the consumer entertainment industry always adapts them in the form of games, toys, movies, and other forms of entertainment. Today we find simulation in a number of computer games and in all animated movies.

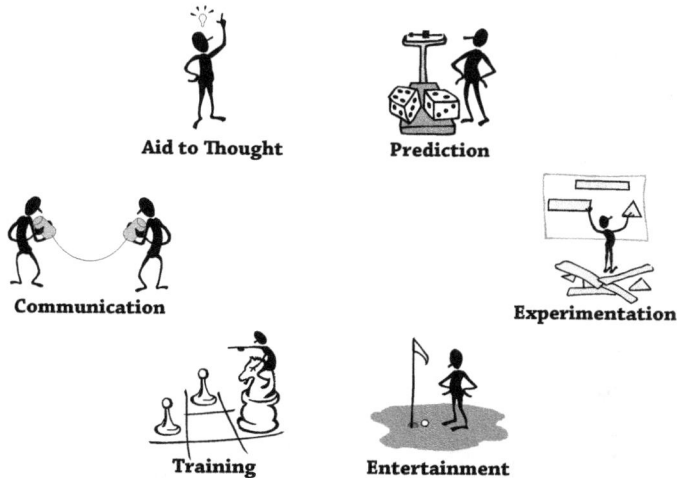

Figure 15.4. Elmaghraby identified five major categories of the application of computer simulation. Entertainment is added as a sixth category.

Though games emerged much later as a useful application of computers and simulation techniques, they have progressed so rapidly that game technologies are now returning to their roots to offer improvements to a number of more serious applications.

Core Game Technologies

In a previous research project, I identified six core technologies upon which games are built, and which offer significant value to serious industries (Figure 15.5). These technologies are:

- 3D Engine for visualizing the virtual world,
- Graphical User Interface for control and management of the simulation/game,
- Physics Models for realistically representing actions and interactions of physical objects,

- Artificial Intelligence for representing the reasoning powers of characters or objects,
- Global Networking to connect multiple players together in a shared virtual world, and
- Persistent Worlds to allow experiences to be cumulative from one session to another and to provide coherency across numerous player experiences.

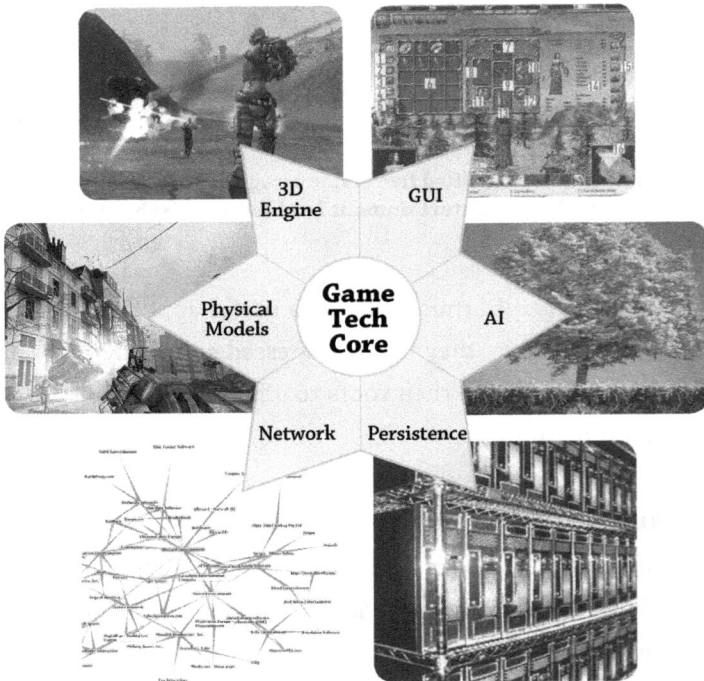

Figure 15.5. Six key technologies that drive computer games

These core technologies play an important role in the spread of games to other industries. Later sections discuss the evolution of serious games and illustrate the importance of each of these in different fields.

Forces for Adoption

Also proposed in a previous paper, are forces that are driving companies to adopt the core technologies listed above (Figure 15.6). These forces are:

- Cost advantage of hardware platforms,
- Sophistication of software applications,
- Social acceptance of game tools,
- Successes in other industries, and
- Innovative experiments in the adopting industry.

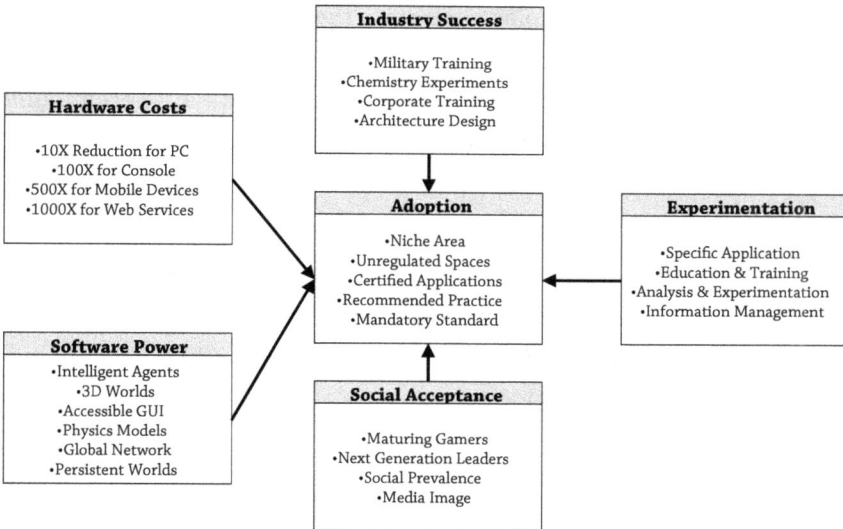

Figure 15.6. Game Impact Theory: Five forces behind the adoption of game technologies by diverse industries

These forces drive experimentation, applications, and perhaps standardization of game technologies in other industries.

Disruption of Established Industries

In this paper we will focus on how core game technologies and the forces for adoption are changing competition in five large industry categories - entertainment, training, scientific analysis, decision making, and marketing. There have been a number of technologies that have significantly changed many industries in the last two decades, including computers, the Internet, business information technologies, and cellular communications. The changes wrought by these may provide clues to changes that lie ahead in the adoption of game technologies. Some specific questions of interest are:

- How will game technology grow into these five application areas?
- How will it change current industry balances?
- Do the impact patterns of computers, the Internet, telecommunications, and IT indicate a similar trajectory for game technology?
- What is the gap between where game technology is now and where it will be at full adoption?

Following in the steps of the dot.com boom, the game industry presents a powerful value proposition for adopting them. Bergeron (2006) identified five benefits of adopting game technologies in serious industries:

- **Economics.** These tools have lower costs, broader dissemination potential, and wider accessibility.
- **Educational Effectiveness.** Games have the ability to illustrate ideas clearly and to communicate to large audiences. Games are an infinitely patient tutor that can repeat material until the student grasps it.
- **Efficacy.** They connect with students through visual, auditory, tactile, and emotional stimuli. Combined, these are more effective in games that attempt to change behaviors or change implicit self-image.

- ***Emergent Behavior.*** Sufficiently rich games create an environment with multiple logical paths and unforeseen advantages to player co-operation. These allow emergent and unpredicted behavior from the players.
- ***Social Impact.*** Multiplayer games create a social space in which players create a community with standards, mores, ethics, relationships, and extensive conversation. These social spaces transcend the content of the game (also cited in Steinkheuler, 2003).

The progress of games for entertainment is driven by short-term market forces. In all of the six core technologies above, game development companies are continually equipped with technological advances such as new graphics cards, physics models, artificial intelligence, and network connections. They are motivated to incorporate these technologies into products very rapidly—specifically, in time for the shipment of a new game every two years. Therefore, research in the game industry is very sales driven. If a new technology can provide a market advantage it is incorporated, if it cannot then it is ignored. This type of research was labeled by Donald Stokes (1997) as "Pure Applied Research". The historical icon that Stokes used to typify this was Thomas Edison who created new technologies solely to create new products. Stokes contrasts this practice with the work of Neils Bohr who conducted research purely to understand the forces of the universe and with no thought of the market potential. This he labeled "Pure Basic Research". Straddling these two categories, Stokes identified Louis Pasteur as a "Use-Inspired Basic Researcher". Pasteur explored the unknown, but with an eye toward knowledge that could be applied to benefit society (Figure 15.7).

Consideration of Use?

		No	Yes
Quest for Fundamental Understanding?	Yes	**Pure Basic Research** **Academic Studies:** • Psychology of learning • Knowledge transfer • Group dynamics • Role of components	**Use-Inspired Basic Research** **Serious Games:** • Validated applications of new technology • Parallel exploration in research and application
	No		**Pure Applied Research** **Entertainment Games:** • Apply new technologies immediately • Make it work • Focus on market potential

Figure 15.7. Entertainment game research as it falls into the research quadrants proposed by Stokes

In the games space, entertainment games follow the research path of Thomas Edison, always seeking to turn technology into an immediate product. University researchers in education, learning, and sociology may follow Bohr's path to understanding the forces behind games for the sake of the knowledge itself. But, serious games practitioners follow Pasteur. They require a validated foundation of the impact of games before implementing that knowledge in a product. The games they create have a market benefit, but must be based on theories that can be substantiated as well.

Using the works summarized in the tables and figures above, we have identified five major categories for the application of game technologies. These are:

• Entertainment,
• Training,
• Scientific Analysis,

- Decision Making, and
- Marketing

Establishing and specifying these categories will be the focus of the rest of this paper.

ENTERTAINMENT

Computer games began as research experiments in university labs. The first is considered to have been "Spacewar!" developed by scientists at MIT on a DEC PDP-11 computer (Figure 15.8). That game represented two spaceships that fire small missiles at each other. Each ship was controlled by a joystick input device and displayed on a circular screen. The similarities with the arcade game of Asteroids are very clear.

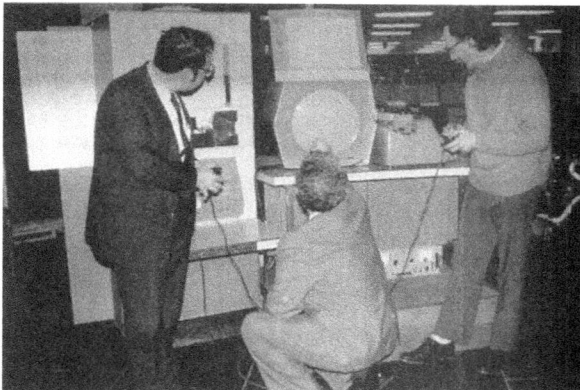

Figure 15.8. Researchers at MIT working with the first computer games—Spacewar!
Source: Michael 2006

Gaming Arcades vs. Pinball & Shopping

Commercially the first arcade game was PONG which was delivered to pinball arcades, pool halls, and bowling alleys in 1972. Developed by Nolan Bushnell and distributed through his newly formed Atari Corp., this game began the transformation of analog pinball games into digital arcade games. At the same time, Magnavox released the Odyssey home gaming system. This included the game of PONG which they claimed Bushnell had copied and Magnavox filed a law suit claiming patent infringement (Sheff, 1999 and DeMaria & Wilson, 2004).

In 1972, the digital computer game moved into pinball arcades and into people's homes simultaneously. The popularity of the new devices created a rush of imitators and began the growth of game arcades and home entertainment consoles. DeMaria and Wilson (2004) provide a clear history of the emergence of arcade games. In this paper we are most interested in the business disruption impacts of those games.

Pinball machines had been in existence since 1871 and a number of companies had been founded upon them. Pacific Amusements added electricity and a ringing bell in 1933 and more complex electrical bumpers, gates, and flippers began to appear in 1947. These devices continued to evolve, adding minor computerization in the 1970's. But their form was largely fixed by then (DeMaria and Wilson, 2004).

Computer arcade games like PONG eliminated the metal ball and the use of gravity as a primary power source. They opened the door to create many new forms of games incorporating sports themes like ping pong, tennis, and handball; space themes like Space Invaders, Asteroids, and Defender; and abstract themes like PacMan, Tempest, and Tetris. These games introduced computer graphics, human interfaces, and audio to the general populace and to the entertainment business (Bushnell, 1993).

Arcades created a casual part-time customer. They could be used to attract crowds to businesses like bowling alleys, restaurants, and malls. In the latter case, arcades added a child attraction to the shopping mall experience that was geared around adults. In a survey of the connection between entertainment and shopping behaviors in 500 large malls, Kim et al (2005) determined that "a good amusement area for children" was a desirable feature in the mall environment. However, those authors did not specifically investigate the role of the game arcade. Bloch et al (1994) report that 21% of people in a mall are there to "consume services" such as movie theaters and game arcades. But only 7.2% report that they have played a video game at the mall. Compared to other non-shopping activities carried out in the mall environment, this is relatively low and suggests that games do not contribute significantly to mall revenues. But games are part of the "habitat" described by Bloch that encourages people to spend their time in the mall. A habitat is an environment which meets people's needs, effectively creating an alternative to their home. Both Barnes (2005) and Bloch et al (1994) admit that relatively little research is done specifically on mall shopping behavior. Therefore, it may not be possible to quantify the impact that game arcades have had on behavior and shopping in these environments.

Home Consoles vs. Television

Consoles like the Magnavox Odyssey and the Atari 2600 moved the casual arcade game into the home. This meant that customers could play the games as often as they liked and without feeding a constant stream of quarters into a machine. However, rather than replacing the demand for the larger, more powerful, and more colorful arcade versions, they seem to have increased the appetites of customers for these more advanced arcade games and for the social environment that was growing up around the arcades.

Home console gaming became a popular leisure activity like watching television. It was always accessible, required little physical activity, and could be played alone or with other people. This began the current 35 year incursion of gaming into the space previous held almost solely by the television. As gaming has grown more popular, this has become a serious threat to the advertising revenue that drives television programming (Bushnell, 1996).

Home PC vs. Web Surfing

Games on the home PC extend entertainment to a machine that was previously purchased for "serious" applications like online communications, banking, and education. In moving to this machine, games became instantly accessible to professionals who may not have been inclined to purchase a game-specific console. This move leverages the already developed expertise that the audience has with the PC to smoothly transition them into gaming.

Each move to a new venue or a new device extends the circle of users and the time spent on games. The trend in the game industry is to find an opportunity for people to access computer games any time and any place, following similar previous patterns for cellular communications, computing, and Internet access.

Portable Handheld Gaming vs. Radio and Digital Music

The introduction of portable handheld gaming devices like the Gameboy series (Standard, Color, Advance, SP, DS, and DS Lite) and the Playstation Portable (PSP) moved gaming from a fixed site activity to one that is portable and accessible in any environment. These devices are to casual activity what the Blackberry and Treo are to mobile business activities. They displace other activities that filled spaces while traveling or away from a desktop computer. In business, mobile computing and communication devices encouraged people to be constantly connected to the electronic web of relationships. They no longer use travel and waiting time to read, think, or relax. Instead this enables them to do traditional office work in the mobile environment. Portable gaming devices extend electronic entertainment to these same spaces. They allow the displacement of previous non-digital activities with "always on" gaming.

The newest Gameboy DS device also contains an 802.11 wireless chip that allows a player to interact with others who are playing Gameboy DS titles anywhere in the world. This means that multiplayer gaming has also moved to the mobile spaces.

Cellular Gaming vs. Cellular Communication

The success of cellular telephones has been driven by an increasingly busy and mobile lifestyle and by their ability to sustain networks of relationships via voice conversations and text messaging. As these devices have become more powerful, their larger color screens, computer processors, and network bandwidth have made them attractive devices for gaming applications. Shuster (2003) estimates that games for cellular phones is approximately a $350 million business. Manufacturers and service providers see gaming as a means of selling devices and advanced services. Just as people once upgraded their computers to be able to play the newest games, the cellular providers are hoping that games will drive this trend in their space as well.

In many ways, cellular gaming is socially similar to portable game devices. However, the major difference is in the ubiquity of cellular phones and the access to an audience that is not limited to those who invest in a game-specific device. The trend is similar to the move from home game consoles to the home PC—both allow the gaming industry to access people who purchase an electronic device primarily for serious purposes, but who are willing to add games to that device.

Hybrid Devices vs. Physical Activity

Interesting combinations of devices are being created to explore a less passive application for games. The largest category of hybrid devices is in exercise equipment, or "exergames" (Michael, 2006). The first step was when game arcades introduced dancing games that required a player to stand on a platform and control the game with very basic dance moves. The success of these devices in Japan spread to America and Europe and demonstrated that physical activity in conjunction with gaming was acceptable to the customer base. Games like *Dance Dance Revolution* have been very successful titles in the arcades and on home consoles, and have driven the sale of peripherals like the electronic floor mats that are the controllers for game consoles.

The Eye Toy for Playstation 2 uses a camera as an input device. Given the right lighting and background, the camera can identify the movements of a player in its field of view. Those movements are then applied to an in-game character to kick a soccer ball, slap balloons into the air, or knock down plates and other objects in the virtual world.

Specialty exercise machines have been created for the game consoles and for home PCs. These devices often mimic the movements of a treadmill, stair climber, rowing, bicycle, or cross country skiing machine. Working with the exercise equipment causes the scene in the game to change so that an exercise bicyclist may experience a virtual ride through the countryside. Hills in the virtual world can be translated into increased pedal resistance on the exercise equipment and turning to the left or right with the bicycle causes the virtual rider to veer off of the trail and into the woods (Michael, 2006).

Using deer hunting games on the PC as a departure point, one company even linked a real rifle to a computer on the Internet. After paying a "hunting fee", a player on his or her home PC is given control of the real rifle on a computer driven aiming platform (Figure 15.9). When a real deer enters the area, the virtual hunter can aim and fire the rifle from their home PC. If the real deer is killed, the virtual hunter has the option of paying to have the animal's meat prepared or its head taxidermied and mounted, both of which are shipped to the real home of the virtual hunter (BBC, 2004).

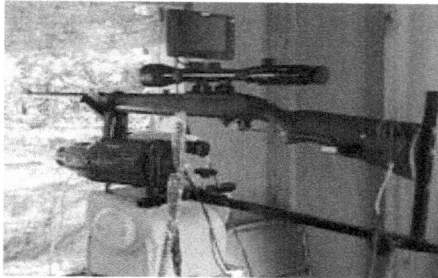

Figure 15.9. Live Shot Internet computerized deer hunting device

Each of these devices demonstrates the penetration of computer games into activities that traditionally include physical exertion. Games are becoming a ubiquitous part of society and our lifestyles. Just as computer scientists predicted that their hardware would become ubiquitous in society, it appears that gaming applications are a significant part of the reason that people adopt new computer equipment.

Content Spin-offs vs. Movies, Magazines, and Books

The popularity of gaming has created a number of spin-off businesses. Each gaming platform has at least one magazine devoted to exploring the hardware and the games that are available for it. There are also books dedicated to teaching people how to play the game better and how to use the "cheat codes" that will give them special capabilities. More recently, the most popular games have created characters and worlds that are the subjects of an entire series of science fiction or fantasy books. Game images drive the sale of logo merchandise that includes clothing, backpacks, book covers, and office supplies. Games like Tomb Raider (Laura Croft), Doom, Final Fantasy, Resident Evil, and Wing Commander have been turned into feature movies. At this point each has had very limited success and is generally reviewed poorly. But they are testing the ability to turn game properties into movie properties. Given the success of the movie adaptations of comic book heroes like Superman, Batman, and the X-Men, it is just a matter of time before they find the right mixture of theater and game characters to achieve the same success.

Game publishers are also signing two-way advertising deals in which game characters are used to promote products like soft drinks and candy, and in which product advertisements are placed in the content of the game.

Combining games, artistry, animation, and moviemaking techniques, a small community has invented a new entertainment form called "machinima", short for machine animation. These are movies in which all of the visuals are created by capturing the imagery from a computer game. Various scenes are then stitched together to create a continuous movie, background music is added, and actors add voices to complete the story. These 2 to 10 minute movies are distributed via the web in standard video formats (e.g. mpeg and avi). The company Chicken's Teeth Inc. has created a very popular multi-year series called "Red vs. Blue" using Microsoft's Halo game as a movie lot (Carroll and Cameron, 2005). Lionhead Studios has also created a game simply titled "The Movies" which extends basic machinima techniques to the general gaming audience.

Like the characters of history and religion, the characters and storylines of games are becoming part of many cultures in the world. These characters will continue to grow to play a larger role, but in this case the characters transcend a number of nations and cultures. Legends and characters from games are not limited to a single race, region, religion, or nationality. This will create a significant change in the separation and unification of national cultures in the future.

Technology Insertion

Many industries are adopting game technology to other forms of entertainment, but without keeping specific game characters or storylines. Television weather reports are now animated with game-like tools. They are no longer limited to the simple two-dimensional artwork of just ten years ago. The weather is now animated and three-dimensional. Similarly, sportscasts can be enhanced through the use of games like Madden NFL and College Day. These games can portray specific teams and even specific players in near-perfect renderings. They allow a sportscast to recreate specific plays, examine them from a number of angles, and even create variations that cannot be accomplished by replaying film footage. Major League Baseball commissioned a game development team to recreate entire real world baseball games using the tools and environment of Second Life (Carella, 2006).

Experiments are underway to cross print, television, and gaming advertising. Promotions are being created to move the advertising audience from a television commercial, to the purchase of a magazine, to visiting a web page, to entering a game for points or clues that allow people to compete for prizes. These treasure hunt stories have been used for years, even going back to radio advertising, but their extension to game environments is new and untested.

Figure 15.10 summarizes the growth of games and their impact on a number of previously established entertainment formats.

Pinball &
Shopping

Exercising &
Hunting

Television
Viewing

Arcade
Machines

Hybrid
Devices

Game
Consoles

**Game
Core**

Cell
Games

PC
Games

Telephone &
Text Messaging

Portable
Devices

E-mail &
Web Surfing

Radio

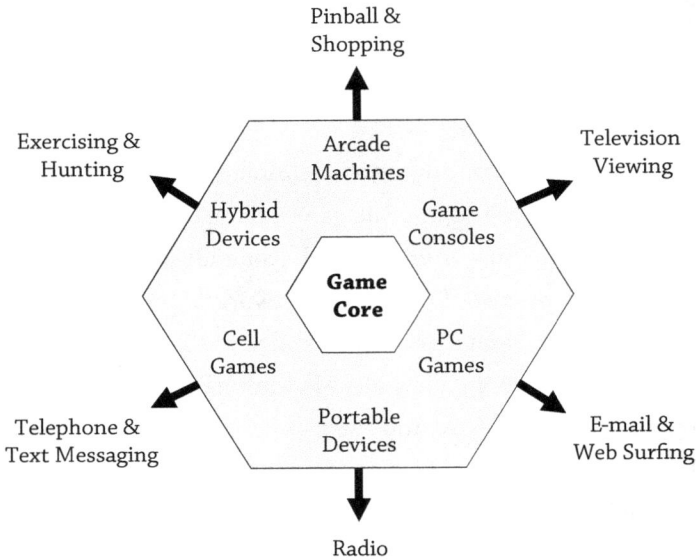

**Figure 15.10. As games appear on a new electronic device,
they compete with established activities**

TRAINING

Training is an essential function within all organizations. Distinct from education, training focuses on conveying or mastering specific skills that will be used for specific jobs by specific people. Public and university education prepare people to learn to work in a profession, providing more of a broad foundation of knowledge. Specific applications require unique knowledge and skills, often unique to a single company or job position. Conveying the necessary knowledge is an expense to the organization in both time and money. The cost of instructors, instructional materials, classrooms, and student time can be significant. Games appear to be able to reduce the costs of training and to improve its effectiveness in some cases.

Bloom (1984) has shown that one-on-one tutoring can improve a student's test scores by as much as two standard deviations. This can effectively move a student from the 50th percentile of a class to the 96th percentile. Though tutoring is extremely effective, it is also very expensive and time consuming. Even if there were no financial costs for a tutor, there are not enough domain experts available to serve all of the students seeking instruction. Because of the costs and limited availability, self-directed study has always been an option to formal classroom or tutored study. Correspondence learning requires that the student learn at his/her own location and often on his/her own time. This extends learning to a much broader set of students and it reduces the cost of delivering the training. Computer-based training (CBT) and web-based training (WBT) have become a very large business because of the significant cost savings involved. Computer games offer a richer environment in which to host computer-based training and student collaboration.

Like all learning materials, games attempt to communicate information in a form that is accessible to the student and that motivates the person to learn. A study by The Learning Federation (2006) identified a number of themes that are driving the training business to a game-based solution. These are:

- **Accessibility.** Games have lower costs for materials and staffing.
- **Believability.** Games present a 3D environment that is so similar to the real world that lessons fit well into their natural environmental context and are more believable.
- **Engagement.** Games can interact with the student in more ways than simple online tests. These engagements keep the students' attention and increase their learning experience.
- **Accuracy.** Given a richer environment, games can create a more accurate representation of the real situation than can materials like textbooks and 2D images.

- **Variability.** Games can present situations in many different forms and in various contexts to reinforce lessons and to show special cases.
- **Measurability.** Games allow the numeric collection of performance data. These measurements enable scoring and identification of the tasks that the student has not mastered.
- **Repeatable learning loop.** Games are infinitely patient. They can take a student through identical or similar scenarios as many times as it takes for them to learn the lessons.

Though the advantages described above do exist, there remains some question as to the effectiveness of game-based training. Like all other forms of training before it, we expect this method to be less effective that real "on the job training", but we do not know how it compares to older forms of training. A recent study by the Federation of American Scientists emphasized that games **may** develop higher order thinking, but they were careful not to assume that this was true in spite of potential advantages like those listed above (FAS 2006a).

In another document generated by that same study, the Federation of American Scientists identified some specific learning advantages of games (FAS, 2006b):

- **Contextual bridging.** Students are able to apply lessons from 3D games much more readily to the real world because of the similarity of the two environments.
- **High time-on-task.** Because of the high level of immersion and inter-activity, students spend a significantly larger amount of time on tasks in a game environment than with other learning materials.
- **Motivation and goal orientation, even after failure.** Students are highly motivated to complete tasks successfully. Failure in game situ-

ations does not seem to discourage students, but encourages them to try different tactics.

- **Providing learners with cues, hints, and partial solutions to keep them progressing.** Game environments can subtly provide students with hints on how to perform tasks. Like a good tutor they can be programmed to recognize when a student is choosing an incorrect path and give hints as to why they should change direction or tactics.
- **Personalization of learning.** Games allow the student to personalize their avatar in the game. This customization builds a stronger relationship between the student and the materials. Given some basic information, the game can also address the student by his/her preferred name and make references to other personal interests.
- **Infinite patience.** Games do not become irritated or impatient with students who learn at different rates. The game does not insist that people learn within a specific number of iterations or in a specific order. They present an ever-present environment that can be explored.

The Learning Federation has observed that the "expense and related challenges often cause both formal education and corporate training to rely on strategies that ignore the findings of learning research" (TLF, 2006). Like computer-based training before it, games have the potential to bring additional learning techniques into an affordable range for industry, government, and academia. Moving more of the knowledge and skills of human instructors and tutors into an interactive, self-directed environment like a game can potentially extend this limited resource to a much higher number of students and spread the costs of the tools broadly enough to make them affordable.

However, TLF also observed that the training and education market is highly fractured. Public schools are divided into independently funded counties, companies structure their own internal training, and government spending is divided into thousands of small offices. They concluded that this audience is divided into too many small pieces to be able to afford the high development costs of building a game from scratch. Their recommendation is that federal funding be used to create the foundation for many of these potential customers. However, in contrast to this perspective, an entire "serious games" industry is springing up to leverage game engines, tools, and artwork that have already been created for entertainment. Their approach is to license these after the development costs have been recovered through the marketing of a successful game. These companies operate with very limited budgets, often encouraged by the generous licensing arrangements from the game product owners. Many of the major game development companies allow smaller companies to use the tools for free up to the point that they create a commercial title. This means that small studios can build their expertise, create demonstrations, and win a client before paying for licenses.

This approach has been working very well for companies focused on a wide variety of customers. In the training space, successful projects have been created for the following types of customers:

- **Military**. Arguably, the military has been using games for training since 1664 when Christopher Weikenman introduced Kreikspeil to the German Army (Perla, 1990). The American Air Corps purchased a number of flight trainers from Edwin Link in 1930, devices which were also being sold to game arcades. The adoption of computer games is a very natural step for this customer, especially since many of their problems are fit well into a 3D virtual world.

- **Emergency Management**. There are a number of new game-based trainers for emergency management that present firefighters, medics, policemen, ambulance drivers, and resource managers with situations that require expertise that is not readily gained through daily operations (Musgrove 2006, Hamm 2006, Rowland, 2006).

- **Government**. Management of public lands, ecological decisions, public policy consequences, and ethics training have all been incorporated into a game form for various branches of the government.

- **Education**. Key events in history are often based on geographic location, equipment, and key people, all of which can be compellingly represented in a game. Physics, biology, chemistry, and business are other subjects that have worked well within a game. The communications that occur in massively multiplayer games has been shown to encourage and enhance reading and writing skills (Steinkuehler, 2005). Games are also an effective environment for presenting logical problem solving.

- **Corporate Training**. In the corporate environment games are an economic way to teach ethics, company policy, customer service, safety, and organizational structures.

- **Religion**. All major religious leaders have used stories to convey the principles of their beliefs. These stories have been passed on through oral and written media for generations. Games are just another tool for capturing, visualizing, and communicating these value-based stories.

- **Health Care**. Games are entering the health care industry almost as fast as they entered the military. They are being used to educate a wide variety of personnel and even to achieve certification. There have been arguments that games are not sufficiently detailed to train a surgeon, but the real benefit is in using games for the hundreds of functions that are carried out in the health care profession which do not involve working directly with human patients. Basic skills, standards, and behaviors can be conveyed to a large number of people distributed across many locations. Games are also being used to treat mental disorders like phobias. They allow a person to be in a simulated threatening situation, but with enough detachment and control to avoid triggering a hysterical reaction.

- **Airlines**. Like the military, the airlines have always used computer simulation to train pilots. These devices allow pilots to rehearse potentially lethal situations in a controlled and learning environment. Games are just a miniaturization of the many simulators already used by the airlines.

- **Safety**. Safety training is conducted in most professions. This training is usually limited to classroom discussion, slide presentations, and printed reading materials. Most people never have an opportunity to act out the steps they are expected to follow in a real emergency. Games create an environment in which everyone can walk through a simulated emergency and play a number of different roles.

SCIENTIFIC ANALYSIS

The 3D engine that creates the visualization for games is a powerful graphic processor known generically as a scene-graph. The images that it produces are simply game specific versions of 3D scenes that are created in a number of different scientific disciplines. Computer Aided Design (CAD) tools allow engineers to create digital models of physical products like automobiles, aircraft, and kitchen appliances. Movie animation tools do the same for computer generated movie characters like those introduced in Pixar's Toy Story and the evolutions that have followed. In games, CAD, and animated movies there are two essential tools for creating the 3D world. Modeling tools are used to design static 3D representations of physical objects. Graphics engines or scene-graphs animate the static models by driving interactions between the objects and with the environment.

Data Visualization

National laboratories employ supercomputers to analyze the behavior of nuclear explosions, geologic movements, weather patterns, astrophysics, and other complex phenomena. These analyses typically begin with a numerical model that represents the movement and interactions of thousands or millions of subcomponents of an item or event. These computations generate extremely large volumes of numerical data that must be analyzed mathematically and statistically. Potentially, all of the information in this data can be extracted numerically. However, in practice scientists have found that visualization is an extremely powerful tool for understanding the behavior of the data and looking for anomalies in the models that created the data.

Robert Rosner, the Director of Argonne National Laboratory, explained that he was originally skeptical of the real value of visualization in the work of the energy labs. However, in one case an Argonne project used a giant, high-resolution display to plot every single data point that had been calculated by their models. Prior to that, much smaller plots were created from a thinned or averaged data set. But when every point was visualized, the scientists noticed a pattern in which aberrant behavior appeared at repeating thresholds in their equations. This visually exposed inaccuracies in the model which were too subtle to be readily identified statistically. Understanding the importance of this, the director became a champion of data visualization and now supports its use whenever possible (Rosner, 2006).

Current game engines lack the fidelity to present the high resolution data that come from many scientific models. However, they remain potentially valuable for rapid data visualization before in-depth analyses are conducted.

Graphing and Animation

Animations of solar systems, molecular arrangements, and chemical reactions represent small scale problems for which game engines can be readily applied. These tools present significant value in lowering the price of animation and in extending access to the tools due to their relative ease of use compared to their competitors. Commercial quality game engines can be licensed for a few hundred dollars, where more traditional visualization tools cost many thousands of dollars for a similar license. Additionally, game engines are designed with an "application programming interface" that allows programmers to attach new software to the game engine to drive it. This means that the game engines can potentially be connected directly to scientific software that is generating data to be visualized.

Presentation

Both traditional science experiments and the more recent computational science contain information that is often difficult to grasp or to communicate. 3D visualization engines are a means for communicating this information to heterogeneous audiences. Rather than creating traditional "science museum" plaster models of human organs, a game engine can create a moving 3D visualization of the inside of the body at work. It can also add the sounds that should be present in such a scene.

These visualizations can be portrayed on desktop computer screens, embedded in head-mounted displays, or projected onto giant walls for viewing by and immersion of an entire audience.

Immersive Navigation

Science fiction has presented stories in which military, medical, scientific, and maintenance personnel navigate a real space by immersing themselves into a virtual representation of that same environment. This technique has become reality in many modern systems.

Maintenance personnel for the Boeing 777 aircraft have access to a virtual overlay repair manual. The repair person dons see-through display glasses and approaches the aircraft. Projected on the glasses is a schematic of the aircraft engine that is superimposed over the real engine in front of him. The schematic may include labels to identify all of the parts and may color code the part that is to be worked on. When it comes time to do the operation, the schematic may even show an animated instruction sequence that indicates the best method of performing the operation. Potentially this can allow a repair person to work on a specific problem for which he has not received prior in-class training. In military environments this can allow a skeleton crew to work on a wide variety of equipment by learning specific information just as they need to use it (Boswell, 1998).

Similar displays are being used to position radiation sources around a patient who is about to receive treatment. A 3D overlay augments a technician's native ability to visualize the location of the tumor and the exact angle of alignment for equipment. The visualization also has the power to identify dangerous situations and to alert the technician to check for these before proceeding with treatment.

In these cases, it is the raw graphics processing power of the software game engine and the hardware graphics card that are desired. Older systems typically had to simplify the displayed information because of limited computing power and limited computer budgets. But as games and graphics cards have become more powerful, they have opened the door to much higher resolution than existing overlay/navigation systems based on older hardware and software technology.

Networked Research

When research is carried out in multiple facilities, a means is needed to exchange working data, leverage laboratory resources at each facility, and collaborate among all parties involved. The new generation of Massively Multiplayer Online Games (MMOGs) are designed to accomplish these types of needs for the entertainment community. An MMOG creates a shared virtual world that can be entered by a large number of participants from around the globe. The virtual world is persistent because actions taken there or objects deposited remain in those locations until acted upon by a participant.

For scientific research, this means that data, or an icon that represents and locates a data set, can be deposited in the virtual world by one player and retrieved, copied, or processed by any number of other players. Currently, Second Life is the environment that is most supportive of this type of operation. It allows participants to load graphics, textures, documents, presentations, and a number of digital data forms into the world. These can be viewed and manipulated in the world or downloaded to a local computer for manipulation (Figure 15.11).

Figure 15.11. Microsoft PowerPoint presentation being given on a virtual projection screen inside of Second Life

Social structures have evolved within all MMOGs. Players have created teams, guilds, societies, and alliances that bring them together for a common purpose. These groups hold regular meetings to discuss their missions and to manage the operations of the group. Many of these are run like business operations, even to holding weekly teleconferences with participants to conduct group business and plan future quests (Steinkuehler, 2007).

Second Life has become a popular venue for holding university classes, distributing assignments, and presenting work completed. It can be used in the same way to bring together researchers for collaboration or to partition work out to a number of remote participants.

DECISION MAKING

Leaders in business, government, academia, military, and non-profit organizations are faced with the necessity of making large numbers of decisions every day. The study of decision making and techniques employed in it is quite diverse. In general, people are guided by principles, objectives, and information. Establishing principles and objectives focuses a decision, but evaluating all information available tends to diffuse the process. Also, many decisions involve multiple parties with different objectives and principles. Therefore, decisions often must be compromises designed to partially meet the objectives of each party, rather than optimizing for any one party (e.g. Janis, 1989; Jennings and Wattman, 1998). Specific processes and tools for decision making will be explored in a separate paper. In this paper we will restrict our discussion to the contributions that game technologies can make to this process.

Historically, decision makers have turned to advisors and reliable sources of information to aid them. More recently, computers have been employed to organize large volumes of information and to present multiple perspectives on a problem. Within the field of artificial intelligence, computers are programmed to provide advice. These programs attempt to codify a body of knowledge and process it according to specified rules. The value of the advice generated is proportional to the degree to which real world information and considerations can be encoded into a form that can be processed by a computer. However, limits in technology, time, and funds available to create such systems impacts their ability to provide actionable advice. These limitations apply to games and decision making systems derived from them as well.

Reasoning

Computer games contain simplified artificial intelligence algorithms that enable computer controlled characters to act somewhat intelligently in a very restricted environment. Typically, the AI is custom designed for the physical space in which the action is taking place and with prior knowledge of all of the relevant objects that will be involved. Game reasoning is not general purpose or universal. It attempts to mimic intelligence in a limited space, not be intelligent in a number of different circumstances.

Understanding this limitation indicates that, to be effective, a game may need to be based on a knowledge base of prior situations that were very similar to the one in which a current decision maker is involved. Because game reasoning is so customized, it may not be appropriate for unique new situations for which prior data is not available. Game reasoning is not the only domain limited in this way, medical expert systems designed to assist doctors in making a diagnosis are based upon a knowledge base of a very large number of previous, similar situations as well.

Illustration and Demonstration

Games can provide a valuable environmental context into which data is injected from other programs. Just as maps are used to provide background and context for reports on the movement of military units on the battlefield, creating a realistic environment in which to portray facts that are relevant to a decision maker can make valuable contributions to understanding those facts.

Games for entertainment often rely upon this context to convey more realism than actually exists in the models that underlie the graphics. Players assume that each visual detail represents a model capability, when this is far from the case. For example, players see that all ground vehicles stick to the surface of the earth, while helicopters fall to the ground when shot. They assume that the game includes a model of the gravitational forces of the earth to make this happen. In fact, almost no game models gravity, instead they enforce simple rules about the behavior of objects that make it appear as though gravity exists in the model.

This feature is a bonus when the goal is to engage the human mind in considering complex situations. There are so many contributing factors, that it is often impossible to model them all, though many of them may be represented visually.

The animating power of games can also be used to demonstrate the dynamic characteristics of data and to project potential impacts of different decisions. A recorded game session may be shared in the same way that a movie is shared. Setting up a situation and allowing a team of people to experience the same space in which decision makers are working can convey an appreciation for the complexities involved that is not possible with non-dynamic, non-interactive tools.

Quantification and Comparison

As digital environments, games are able to record numerical results of actions taken. This can be processed objectively and remove much of the human bias that creeps into decision making that includes a number of people with varying agendas. There are a number of business games that achieve this. Games like "Mike's Bikes", which is used in a number of business schools, place teams of players into an environment in which they compete against each other to maximize the sale of bicycles and the company share price. These games are formula driven, quantifying every decision and keeping a record of the results of each. In Mike's Bikes the actions of each team have an impact on the results achieved by the others as well, creating an interactive competitive environment. All teams are developing, marketing, and selling products into a single market with a defined level of demand for the product.

Given the quantification of multiple decision paths, computer systems can be used to compare the results of each and the consequences expected. Games, like business, usually portray a win-lose scenario that result in one or more companies winning a stake while others lose their stake.

Experimentation

Computer programs are ideal for experimentation with multiple options. When the computer can calculate results very rapidly, it is possible to explore dozens or even hundreds of variations on a decision before identifying the most promising combinations. This digital laboratory is rapid and private. It allows excursion without sharing ideas so widely that they become rumors or competitive intelligence.

Games contribute to this in that they are relatively inexpensive to modify, can be limited to a single computer with access restrictions, and display results in a visually memorable and comparable form.

Business Applications

Games and game technology for decision making have been employed in a number of industries to assist with decision making. These include:

- **Architecture.** The design of buildings has transcended paper models. Digital representations of structures include features like stone surfaces, tint of the glass, carpeting, paint schemes, furniture selection and placement, and division of internal spaces. Placing the new building into a virtual representation of the city also illustrates how it will fit into the landscape, what the view will be from each window, how the sun will play across the glass during the day, and what the surrounding city will look like at night. These models allow the architect and the customer to make design decisions long before any physical structure is created.

- **City Planning.** On a larger scale, city planers can use game tools and environments to layout entire cities, to plan freeways, shopping centers, locate police and fire stations, and track data on events throughout the city. The work of Jay Forrester in System Dynamics encouraged city planners to think of their city as a large interacting system and games like SimCity have allowed them to work with these ideas in a visually appealing model.

- **Law.** Important decisions are made in a court of law. Therefore, both defense and prosecution lawyers have applied 3D visualization to crime scenes. These allow juries to see the situation from a number of perspectives and to understand it better.

- **Leadership and Management.** Games can create a challenging environment in which to exercise leadership. Game avatars controlled by artificial intelligence can present leaders with a wide variety of responses to their actions, challenging them to develop alternatives for each.

MARKETING

Many of the applications described above involved convincing people that a specific perspective on a problem was correct. Though some of those are similar to marketing, we did find examples of games being used purely to market products and ideas. Since there was no other objective, such as making a decision or analyzing data, these cases demanded their own category.

Advertising

Games have been employed as advertisements. Web sites like Postopia. com present children with simple games that prominently feature company products and convey messages about those products. In this sense they are exactly like television advertisements. However, they improve on traditional advertising because the game engagement can keep a child connected to the product and its message for many minutes or even hours. This is impossible with television and print ads. This type of advertising via game is more closely aligned with event sponsorships, like the Nike logo that appears on football uniforms. Audiences fix their attention on the game and on the advertisement for hours at a time (Winkler and Buckner, 2006).

Activism

A number of games have been created to make the world aware of situations like starvation, genocide, discrimination, pollution, and conservation. These games create an accessible and interactive story. Like advertisements, the engagement can capture people's attention for significantly longer periods than traditional narration or printed stories. They convert statistics and data on the situation into realistic, interactive events that place the player in the position of the person at risk. This creates an empathy that cannot be matched in many other media. This application of games is not new. Hutchison (1997) reports on a game created by the Red Cross in 1920 to teach children how to reduce the chances of catching diseases.

Examples of some of the new electronic games are:
- Darfur Is Dying: http://www.darfurisdying.com/
- Food Force: http://www.food-force.com/
- A Force More Powerful:
  http://www.aforcemorepowerful.org/game/index.php
- Escape from Woomera: http://escapefromwoomera.com/

Politics

Games have been used to convey political messages. These may be as simple as illustrating the dynamics of voting in Illinois or exploring the conflict between the Israelis and the Palestinians. In some cases, the game is designed to promote a specific point of view. In others it is meant to communicate multiple points of view so that a player can appreciate the difficulty of a situation.

Examples of these include:
- Peace Maker: http://www.peacemakergame.com/
- Howard Dean for Iowa: http://www.deanforamerica.com/
- Take Back Illinois: http://www.takebackillinoisgame.com/
- Under Ash/Under Siege:
  http://www.underash.net/en_download.htm

Religion

Religious games are usually designed to communicate the core message of the religion and to take advantage of the long-duration contact that people put into games. It also animates historical information that can easily become rote or dull to an audience that has heard it many times over.

Examples are:

- Left Behind: http://www.leftbehindgames.com/
- Interactive Parables: http://www.interactiveparables.com/
- Ominous Horizons: http://www.n-lightning.com/ominoushorizons.htm
- Catechumen: http://www.n-lightning.com/catechumen.htm

The five application areas discussed above are summarized in Figure 15.12, with a space left to indicate that future areas of application are likely.

Entertainment	Training	Marketing
Arcade – Casual part-time access to customers. Matched with shopping. Console – Specialized in-home access. Matched with television. Home PC – Mass consumer access. Matched with IT. Portable – Specialized portable access. Matched with radio. Hybrid – Exercise equipment, dancing, hunting, Cellphone – Mass consumer, portable access. Matched with portable telecom. Content Spinoffs – Movies, news, sports, magazines, and web sites.	Emergency Management – Immersion in major disasters and collaborative operations Logistics – Capture complexity and volume of detailed data. Medical – Drill basic skills, standards, and behaviors. Military – Reduce combat attrition by eliminating mistakes. Airlines – Cost effective flight training, accessible aircraft Safety – Industry jobs requiring safety training. Good for scenario-based training	Advertising – Lock in audience attention on a message mixed into a game. Activism – Deliver a serious message that may be outside of normal human experience and difficult to create empathy. Politics – Motivate to take political actions in favor of the point-of-view of the game. Religion – Electronic proselysation and retention of the faithful.
Scientific Analysis	**Decision Making**	**New Sector**
Visualization – Explore complex data with the primary human sense. Graphing and Animation – Numeric and dynamic representations of systems. Presentation – Communicative representations of complex problems. Immersive Navigation – Ability to navigate and search data in a three dimensional context. Networked Research – Tools for connecting multiple researchers together in an effective manner.	Reasoning – Assistance in thinking through a complex problem in a consistent and accurate manner. Illustration & Demonstration – Visualize, Communicate ideas to broad variety of people. Show how a decision will play out. Quantification & Comparison – Measure the impacts of different decisions. Contrast multiple options and identify the preferred. Experimentation – Consider multiple options and variations that are beyond physical experiments and prototypes.	What new applications will emerge as game technologies progress?

Figure 15.12. A core set of game technologies have enabled the extension of games into a number of different industry categories
Source: Created by the author

ISSUES

In this paper we have explored the many ways that games and game technologies are creeping into different industries. As the breadth of this impact becomes clearer, games begin to look less like toys and more like the 21ˢᵗ century manifestation of advanced computer technologies that are useful for a number of purposes. Games are really an extension of the computer/Internet/ Web/IT transformation that has been occurring across all industries over the last twenty years. Games are unique only in that the technology came from entertainment applications, rather than from some other "serious industry". Games are similar to movies, film, and radio, all of which leveraged a huge market-base to create advances in technology that could be applied in a number of non-entertainment industries.

Because of its origins, game technology faces a number of interesting issues in its adoption by other industries.

Serious vs. Entertainment Origins

Scientists, business people, political leaders, and academics are usually devoted to pushing forward the state-of-the-art in their own fields. They study, experiment, publish, and debate new advancements. But, when a useful technology emerges from a completely different domain, it is difficult for them to accept it as being useful in their craft. The perception is often that truly valuable advances are invented inside of the specialty itself, not imported from the outside.

Game technologies carry the additional burden of originating as "toys for children". Organizations like the military have had a difficult time overcoming this fact because of the extremely serious nature of their business. The most effective forces for changing this issue have been the passage of time and the proliferation of games throughout society. As younger leaders are promoted they bring with them an appreciation for the power of game technologies. They are not wed to the way things have been done in the past and are more open to newer ideas, even those from the entertainment industry. This force is supplemented by the fact that game technologies are appearing in all aspects of life. When a game is used to tell the story of starvation and oppression in Darfur, it suddenly becomes a more serious and valuable tool in a number of other industries as well.

Accuracy vs. Appearance

Games create an environment that is visibly and audibly rich. The details presented imply that the games contain much more accurate models and predictive power than they really do. Like movies, they are designed to imply that there is much more to the virtual world than just the thin facades of the buildings. Many people who begin working with games soon discover that the tools are not able to handle complex problems that are significantly different from their original focus.

Professionals in visual simulation, e.g. flight simulators, have dealt with this for decades. They work to present a very realistic world, but to limit the expectations of the pilots who fly in them. All game developers and users are going to have to participate in a dialog that clearly defines what the game can do and what is merely the appearance of a capability.

Extensibility of Tools

Games work very well for problems that can be rendered on a map or in a 3D space. They have been much less effective at capturing the subtleties of problems like financial analysis, organizational communications, and human psychology. Attempts to render information from these domains has often devolved into an artistic form of spreadsheet that looks and acts very much like a traditional spreadsheet. Spreadsheets and textual documents remain an effective way to convey some information. Games have yet to discover an abstraction for this information that is more effective than existing forms of presentation.

Consistency Across Game Environments

In the commercial gaming industry, every title is a competitor against every other title. As a result, there is very little collaboration among game companies and game tools seldom work together. For example, no two game titles have ever been created which are interoperable with each other. It is not possible for an object generated by one game to exist in another game and interact with objects in that game. This is an essential capability for many commercial and government spaces into which games are moving. Military simulations use a number of data standards that allow systems created by one service and development contractor to interact with those created by another. This allows Air Force fighter jets to fly into the same space used by Army ground forces and for the two to interact with each other through combat engagements, communications, and the exchange of resources.

The fundamentally different perspective of the game industry on this issue may make it difficult for them to design systems which can be more interoperable than the existing military systems that they are trying to replace.

At another level of detail, it is important for the model of an event in one game/simulation to be compatible and comparable to that in another. It is not useful for two different games to arrive at two totally different conclusions when given the same problem to represent and study.

RESPONSE ACROSS THE ORGANIZATION

The expansion of game technologies into new industries is creating opportunities for a number of different types of people. The entrepreneur will find opportunities to create new companies, new products, and new services in this space. Given a personal or organizational competency in this area, this person should be able to establish a firm foundation upon which to create business.

Leaders will find it necessary to make strategic decisions about whether their organizations need to be restructured to adopt these technologies and incorporate them into their products and services. They will also have to determine whether to use these tools in their current industry or to apply them as a bridge to move into an adjacent area.

Given corporate movement into this area, managers will have to work within the organization to equip it to use these tools and to overcome the fear that change will generate in the organization. Managers must deal extensively with people and with organizational structures to create the skills and the structures necessary to apply these new tools effectively.

Administrators will have to measure the effectiveness of game technologies, maintain accountability for the products and licenses, and organize the large volumes of data that these tools both consume and generate. Accomplishing this effectively can make a significant contribution to the competitive viability of these new technologies in the organization.

CONCLUSION

Game technologies are proving to be very flexible in adapting to industry needs outside of gaming. Entrepreneurs are launching companies to take advantage of this disruption. Because of their very specific focus on the visualization of environments and problems, games will probably never achieve the level of adoption that has come to the Internet or IT services. But they do represent the application of powerful new hardware and software technologies being created in the 21st century. It may be fair to expect that the spread of game technologies is the first disruptive wave of computer technologies in this century, and that it will be followed by others that will be larger.

As I described in a previous paper (Smith 2006), it is important for companies to continue to evolve, which includes the adoption of new technologies. Companies that refuse to consider the uses of game technologies may very well hinder their ability to understand and appreciate the waves of technology that will follow. For example, within games are important lessons to be learned regarding computer graphics, networking, shared data persistence, analysis of large volumes of information, and interoperability among disparate tools. Companies that do not learn these lessons now may be ill equipped to apply them to non-game-based applications in the future.

In the last few years, games and the technologies embodied in them have proven to be an undeniable force in changing the way many businesses operate. The ability to bring rich visualization and interactivity to every computer desktop adds value to a number of different types of operations. In this paper we have explored the value in entertainment, training, scientific analysis, decision making, and marketing. The adoption of games in these areas is reminiscent of the use of the Internet and Web in the mid 1990's. There remains some question as to the value of doing this and the business model to be applied, but pioneers are pressing the technology into service in hundreds of small ways right now.

Chapter 16

GAME IMPACT THEORY

I n his history of the game company Parker Brothers, Philip Orbanes demonstrates the impact that a number of games had on the social practices of the United States in the late 19th and early 20th centuries (Orbanes, 2004). As new games were invented or imported from Europe, they changed the way the population spent its time and changed the definition of leisure activity. Through his new company, George Parker introduced board games like Dickens, Ivanhoe, Chivalry, and Mansions of Happiness. Each of these was a moderate success, but more important to this study is the fact that they defined leisure entertainment as a gathering of four to six people around a game board with colorful pieces, die, and rules of play. In defining leisure these laid the groundwork for the explosive successes of Monopoly and Sorry which would be introduced much later.

Since people were gathered about the dining room table playing his board games, Parker added card games like Pit, Flinch, and Rook, each designed to make card games socially acceptable, rather than the unsavory tools of gambling that they were considered at that time. This same redefinition of leisure activities has occurred with computer games in recent years. Game companies from Parker Brothers to the present have found that it is important to work within existing social boundaries, but that these are often poorly defined and open to redefinition by the game companies (Orbanes, 2004).

The definition of leisure has continued to occur throughout the 20th century, and has been strongly influenced by the introduction of electronic gaming instruments. Advances in electronics have brought us the radio, television, pinball machines, slot machines, video arcades, home game consoles, and personal computer games. Each of these has become a part of leisure entertainment and each has led to the creation of major corporations. However, the PC game has created products and a set of tools that are spreading far beyond their entertainment roots and are impacting a number of established "serious industries". The technologies from these games are changing the way numerous business activities are being done, from corporate training to scientific data analysis.

COMPUTER GAMING: BUSINESS & ECONOMY

In his widely read history of the Nintendo Corporation, Sheff (1999) describes how the computer technologies of the 1960's were turned into the first arcade and console games of the 1970's. Nolan Bushnell's "Pong" game took the pinball arcades by storm when it was introduced in 1972. It began the transition from electrical-physical games to digital-electronic games. Bushnell's Atari Company soon turned Pong into a console device that could be connected to a television and played in the home. This was the first step in turning passive television viewers into active television "finger athletes". It also kicked off the creation of the video game industry. Companies like Atari, Magnavox, Nintendo, Sega, Sony, and recently Microsoft entered the home and began to compete with television, board games, and other forms of leisure entertainment. Since that entry, traditional board and card games have found it difficult to compete with moving three-dimensional action that shouts, explodes, and laughs in beautiful colors and perfectly rendered images.

From the humble beginnings of Pong, Pac-man, Tetris, and video pinball, we have moved to Super Mario Brothers, Wolfenstein 3D, Command and Conquer, Microsoft Flight, Quake, Unreal Tournament, Half-Life, and World of Warcraft. Each of these has provided more realistic virtual worlds and more interactive environments in which to play. As the Internet grew, these also extended themselves through the telephone and network lines to create competitive play with other people around the planet. Games have enabled shared play with complete strangers and defined that as an acceptable form of entertainment.

Not satisfied with small team interactions involving 2 to 16 players, game companies created a new genre known as the "massively multiplayer online game" in which thousands of people can participate in a shared world and can return to that world day after day to see it grow and change in response to their own actions. Multi-User Dungeons, Ultima Online, Everquest, The Sims Online, Second Life, and World of Warcraft have created many alternate communities where players can build relationships, teams, businesses, economies, and a complete working social structure that parallels the real world. Games have defined what entertainment is, but they are also defining social mores and financial values.

Games as Business

Computer games are a rich and vibrant part of the economy. In recent years, the size of this industry has surpassed that of the movie industry in Hollywood. In 1999 the US game business was estimated to cross the $10 billion mark, compared to Hollywood's $8 billion. It is also a significant driver of the computer hardware and software industries.

[Note: The 2009 estimate is that the game industry is $50 billion annually.]

Nolan Bushnell, the founder of Atari, pointed out that a number of computer technologies that are now considered standard in business computing were actually invented and applied in computer games. These include raster scan monitors, sprites, real-time graphics, graphical user interfaces, three-dimensional graphics, publicly available computing resources, trackballs, joysticks, sound feedback, collaborative computing, and anthropomorphism (Bushnell, 1996).

Much of this game technology is a direct descendent of military simulator research and development, specifically the Simulator Networking (SIMNET) program sponsored by DARPA in the 1980s and 1990s (Miller & Thorpe, 1995; Dodsworth, 1998). As computers became more powerful, game makers adopted the military theme and style of training as soon as possible. Young programmers like John Carmack created one of the first three-dimensional games for the personal computer, Wolfenstein 3D, and released it as shareware in 1992 (Kushner, 2002).

Like the radio, television, computer hardware, and the Internet, computer games have generated entirely new businesses led by the enthusiasts who jumped in early. These people have also created a number of unique business structures, practices, and partnerships and these are impacting established companies in industries that, at first, appear to be far removed from gaming. But in fact, who share a common technological foundation that the game companies are beginning to exploit. In this paper we explore the economic and social impact that the game industry and its associated technologies are having on industries like:

- Entertainment,
- Education,
- Training,
- Scientific Visualization,
- Scientific Analysis, and
- Exploration.

Games for Business

"The second World War exposed many American servicemen and service-women to simulations and games. When they returned to work they began to see the possibilities of learning facts and insights in an experiential way and how that could be used to train staff to keep up with the business expansion, which came post-war."

(Lane, 1995)

Games and simulations are not just found in entertainment. As Lane noted in 1995, the simulators and wargames that were used to train military personnel during World War II opened people's eyes to the potential of games and simulations as tools for learning and improving performance. This idea had been growing in military research communities for at least three hundred years, going back to the military game of "Koenigspiel" which was introduced in 1664 by Christopher Weikhmann of Ulam, Germany (Perla, 1990). But those ideas were relegated to the government and military communities and did not propagate out of that community for nearly 300 years.

Post-war companies introduced a number of business training games and incorporated competitive game playing into traditional classroom courses. But, as an industry and a business in itself, this had little impact on the economy or the structure of multiple industries. Like Parker Brothers' games, it was the use of these in entertainment that really generated interest and sufficient revenues to drive invention and innovation. One of the most notable of these was the creation of the hobby board wargame by Charles Roberts in 1952. Roberts was an Army reservist awaiting his commission and sought to create a tool with which to practice his tactical decision-making skills. This desire led him to organize and improve upon the wargaming tools that had been evolving since 1664. In addition to military training tools, Roberts created the Avalon Hill Company which started the entire hobby wargame industry (Perla, 1990).

Fifty years later, "Video games and all related industries generated 220,000 jobs and US$ 7.2 billion in wages in the US in 2000" (Aoyama & Izushi, 2002). Figure 16.1 illustrates the explosive growth of the game industry since 1994.

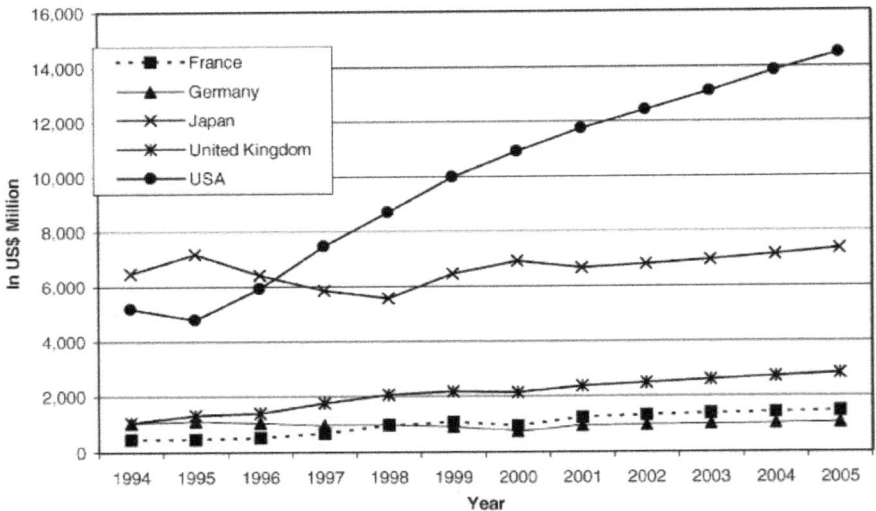

Figure 16.1. Computer game market trends, 1994-2005
Source: Aoyama & Izushi, 2002

Some authors see these games as a very limited and bounded product within a single industry. Maier & Grobler (2000) attempted to group all applications of games to learning into a half-dozen simplified categories. This view ignores the continuously changing nature of games, the computer technologies that support them, and the synergistic relationships between the companies that sell them into every market space. They are much more like malleable tools than specialized applications. Their creators will transform them into any form for which a sufficient customer base exists.

In fact, in an ironic turn of leadership, the game industry is now seen as a core source of innovation, technology, and products to drive the military simulation industry from which it originally sprang. Members of a recent government-sponsored study group on entertainment technologies encouraged the military to establish sharing relationships with computer game developers and the movie special effects communities as a means of pulling these technologies into military training to improve its realism (National Research Council, 1997). This recommendation came to fruition when the Army invested $50 million to create the Institute for Creative Technology at the University of Southern California with exactly this mission (Sieberg, 2001).

KEY GAME TECHNOLOGIES

The power of games stem from the technologies that they contain, their emotional connection to their customers, and the business ecosystems that they have created.

Having evolved from their roots in Pong, Pac-man, Space Invaders, and Tetris, game makers have identified a combination of software technologies that are most effective at winning customers. These have moved from the personal computer, to the television game console, and are entering the cellular telephone as fast as hardware technologies will allow. Figure 16.2 illustrates these technologies with images. These dominant game technologies are:

- 3D Engine
- Accessible GUI
- Physical Models
- Artificial Intelligence
- Networking
- Persistence

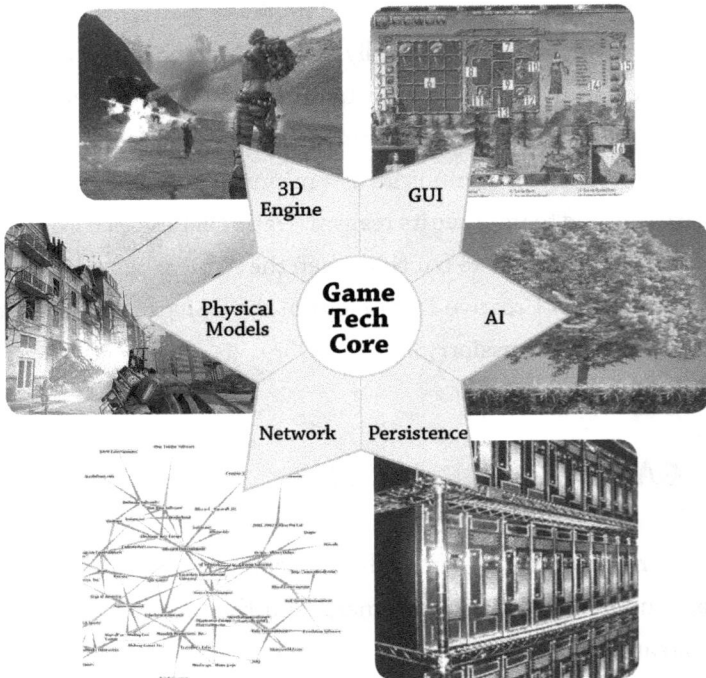

Figure 16.2. Six key technologies that drive computer games

3D Engine

Easily the most identifiable part of a game is the three-dimensional engine that creates the graphics that a player stares at for hours on end. These engines are the key component with which vendors compete for market attention. The detail of the characters, the vibrant colors, and realistic explosions all attract players and commercial users from one previously hot game to the next.

Figure 16.3 provides a side-by-side comparison of the scene generated by Wolfenstein 3D, the first market blockbuster 3D shooting game in 1992, with a comparable scene from the same company's 2005 game entitled Quake 4.

Figure 16.3. Visual comparison of 3D scenes from 1992 and 2005
Source: id Software—http://www.idsoftware.com/.

The 3D engine is not just a game technology. There are a number of industries that require this type of capability to visualize information. Most prominently, these include flight simulator training, medical imaging, architecture, and computer aided design. Smed, Kaukoranta, & Hakonen (2002) illustrate the fact that the virtual environments created by a 3D engine are applied in different ways by different communities (Figure 16.4). Computer games, simulations, and academic virtual environments have significant overlap, but remain unique fields. This picture also illustrates the opportunities that exist for computer games and game technologies to infiltrate these very closely related industries. As we will demonstrate, the number of domains that intersect in this way far exceeds those considered by Smed.

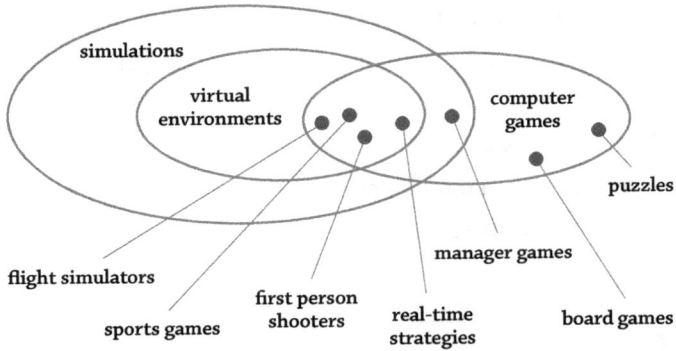

Figure 16.4. Computer games, simulations, and academic virtual environments share a number of characteristics. These commonalities represent market opportunity for the game industry. *Source: Smed, Kaukoranta, & Hakonen, 2002*

GUI

"Why do we put up with the frustrations of everyday objects, with objects that we can't figure out how to use?" (Rosen & Weil, 1995) In their 1995 study, Rosen and Weil discovered that 25% of American adults had never used a personal computer, programmed a VCR, or customized the stations on their car radio. They lay the cause for this squarely at the feet of the complex user interfaces provided by the manufacturer. This is a popular problem for which a simple, non-threatening user interface would find a ready market.

In an attempt to attract and retain as large an audience as possible, computer game developers put significant effort into creating a graphical user interface (GUI) that is easily accessible and understandable. The ultimate goal is to make it possible for a broad audience to play the game without reading a manual. In fact, most games come with a familiarization stage in which the game teaches the player how the controls work and gives them simple tasks to perform to gain and demonstrate proficiency. This expertise in creating an effective human interface can be applied in hundreds of different industries to overcome the issues identified by Rosen and Weil.

Physical Models

When playing a game, characters run, jump, shoot, fall, and explode. Each of these actions is controlled by physical models that estimate how real objects would perform these actions. As games have evolved, the first simple animations of running characters (sprites) have been replaced with physics-based models that calculate how fast a specific vehicle can climb a hill and which objects would impede its progress. The accuracy of these models determines the realism of the world and the pictures that are drawn by the 3D engine.

The most recent advances in this area are the additions of "physics engines" which treat digital objects as if they have mass, weight, volume, stiffness, and joints that operate like real world objects. When movement and interactions occur, the physical properties of these objects interact with each other and with the gravity of the world to generate images that appear to be as real as the physical world we inhabit.

Artificial Intelligence

A game in which only one player exists is very dull. The human player must encounter other characters that are interesting to interact with. These characters possess an artificial intelligence that allows them to perform all of the same actions that a human controlled player would, and to do so in a manner that appears realistic to the human.

The AI techniques created in academic and industry labs have been adopted for games at a very rapid pace. In fact, leading AI researcher John Laird from the University of Michigan commented that the latest advances in AI appear in games long before they show up in more traditional industries (Laird, 2000).

Networking

The Internet connects live players with others around the globe. This means that each person is no longer limited to playing against an artificial opponent on his own computer, but can compete against other people as well. In fact, a player can locate others who have similar skill levels, people who have hours of experience with the game and will play in a unique and unpredictable manner. Like the novels that describe man-hunting-man, this creates the ultimate game of survival in the virtual world.

Networking also enables the creation of a shared community and subgroups referred to as clans. Those who enjoy the games also enjoy sharing stories and tips within this community. In a networked virtual world, there can be much more to the game than just the software in the box. It can open up an entire alternate community in which to live and build relationships.

Persistence

Until 1997, most networked games created small vignettes in which individuals came together in a dungeon, castle, or fortress to team up and fight each other to the death—over and over and over again. But once the characters left the environment, all remnants of their engagement were deleted from the computer servers and the space was refreshed for the next team that would meet there. But, in 1997, the game Ultima Online changed all of that. The creators envisioned a world that was persistent. It exists before a player enters and after a players leaves. The actions that a player takes in the world persist there as they would in the real world. In this environment, the game becomes an evolving story that changes from one day to the next based on what all of the players do there. The success of Ultima Online led to a number of competitors like Lineage, Everquest, Asheron's Call, The Sims Online, and World of Warcraft. These game worlds can support hundreds of thousands of simultaneous players and evolve over many years. An alternate society that is not mundane and in which a player can create a persistent identity is a power-ful attractor for long-term players and long-term customers.

Software Pushes Hardware

These six core software technologies have pushed computer hardware manufacturers to create more powerful equipment. In fact, the game industry has become the primary force driving commercial computer advances – ex-ceeding the demands of business productivity and military applications.

The game community calls for improvements in CPUs, memory, graphics chips and cards, display monitors, network connections, sound generation, user interface devices, and back-end server computers. Advances in all of these benefit all industries that use computers for their products and services, and most significantly, the price/performance ratio enabled by the huge game mar-ket. It also builds an industrial computing environment that is capable of run-ning the game technologies listed above when such an application becomes available in their industry.

GAME POWER: TECHNOLOGICAL, PERSONAL, FINANCIAL, AND SOCIAL

As stated earlier, the computer game industry has risen from near obscurity to a major industry in just over ten years. It has capitalized on a number of forces in technology, society, and individual preference. When brought together, these forces have been powerful enough to propel the growth of this industry faster than more traditional industries at the beginning of the 21st century. Through consolidation, the largest three companies that are purely in the game software business are Electronic Arts, Namco, and Activision. Though these are signifi-cant companies in the game industry, they are dwarfed by the leaders in the industries that are being disrupted by their technologies (Table 16.1). This size disparity demonstrates the huge growth potential that may lie ahead for games.

Table 16.1. Largest companies in select industries (by annual revenue in billions of dollars)

Computer Game Software	Entertainment	Defense	Pharmaceutical	Energy
Electronic Arts $3.1	Time Warner $43.6	Boeing $54.8	Pfizer $51.4	Exxon Mobile $339.9
Namco $1.6	Walt Disney $31.9	United Technologies $42.7	Johnson & Johnson $50.5	Chevron $189.5
Activision $1.4	News Corp $23.8	Lockheed Martin $37.2	Abbott Labs $22.3	ConocoPhillips $166.7

Source: Fortune 500 listing for 2006 and game company web sites.

Computer games harness four different forces in order to achieve their growth and the attention they receive from the population and the media. These are technology, personal, financial, and social.

Technological Power

Computer hardware technologies have exploded in the last two decades. They have moved so rapidly that many companies have not been able to take advantage of the power that they offer. Both the developers of software applications and the business consumers of these have usually not required the constant improvements in power that are available.

However, the gaming industry has moved rapidly to create new software that takes advantage of the latest in computer hardware. They have a relatively rapid software development cycle and a low product price. Since they depend on large numbers of customers to purchase their products, each game can be sold at $20 to $50 and still generate significant profits. The constant demand for new games also drives the industry to create hundreds of new titles every year, with the development time of a single product averaging 18 to 24 months.

Such rapid turn-around allows the industry to experiment with new ideas rapidly and to identify those that work and those that do not. As described for AI earlier, the game industry is often the first place to create a commercial offering of a new software or hardware technology created in a research laboratory.

Thomke (2003) argues that this rapid experimentation with new ideas is an essential part of reducing costs and deployment times. In Figure 16.5 he illustrates that making more mistakes faster can shorten the product development time and improve the quality of the product. Rapid mistakes reduce the costs that accumulate as a project matures, which reduces the cost impacts when a mistake is found. Making mistakes faster literally makes it possible to make more mistakes and to squeeze more problems out of a product before it is deployed to customers. So, in addition to getting to a product faster, it creates one with fewer embedded errors.

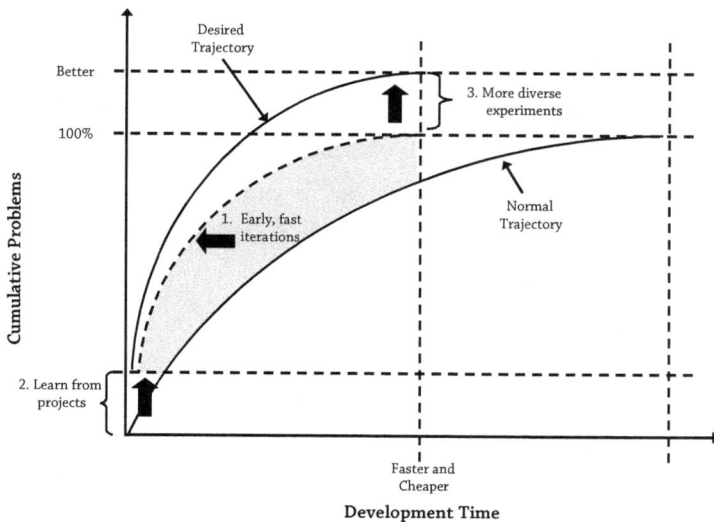

Figure 16.5. Making mistakes faster can reduce product development time and increase the quality of the product
Source: Thomke, 2003

Once they discovered the advantages of rapid product development, the game companies themselves pushed this even faster. They began to market their next breakthrough product earlier and earlier, even to the point that they were promoting products that had not begun development yet. They were locked in a leapfrogging arms race to capture the mindshare of customers, even if they had to sell vaporware to do it (Schilling, 2003).

Personal Power

Eric von Hippel of the MIT Sloan School of Management has written a number of papers on the phenomena in which product users become important innovators of new and more advanced products. He and his collaborators have studied industries as diverse as mountain biking, wind surfing, medical equipment, and open source software. In all of these, there is a community of lead users who demand improvements to the current products and who take it upon themselves to create the improvements that they need. This behavior existed long before the current information economy, going back at least to the agricultural era in which farmers had to make most of the equipment they needed because there was little organized industry available to provide it. However, the difference today is that an organized industry does exist everywhere that lead users are innovating. Also, given the current information connectedness of society, the innovations created by one lead user can be shared and duplicated by others around the globe. This network of lead users can make up an alternative, and competitive, source of products (von Hippel, 2005).

Given this environment, many industries attempt to include these lead users in their own internal innovation process. Their goal is to harness the personal energy of these innovators. The game development community exhibits these same traits of user innovation. An entire "modding" community has sprung up which creates new artwork, new game levels, and new software that can be incorporated into a popular game. The game makers have encouraged this trend by making their games more modifiable with each succeeding generation. The modifications by their lead user communities enhance the value of the core product and drives additional sales.

Shankar and Bayus (2002) point out that the market position and growth of a product are derived from at least two equally powerful forces. The first is the size of the installed base. In the case of console games, Nintendo became a powerful player because the large number of installed consoles drove large numbers of game cartridge sales. However, it was overtaken by both Sega and Sony at different times, largely due to the strength of their network of users, which is the second of these two forces. "Strength" refers to the degree to which customers build their own network around a product, discuss it, promote it, and make modifications to it. The PC game community recognizes this value and actively promotes is customer/modder base. Blizzard Entertainment has produced three major blockbuster games for multiple players – Starcraft, Diablo II, and World of Warcraft. Writers have commented that, "Blizzard's products are computer games, but the social dynamics of a networked player population are the backbone of its business" (Herz & Macedonia, 2001).

von Hippel listed three conditions under which lead user innovation is most likely to grow. These are: "(1) at least some users have sufficient incentive to innovate; (2) at least some users have an incentive to voluntarily reveal their innovations and the means to do so, and (3) user-led diffusion of innovations can compete with commercial production and distribution" (von Hippel, 2001). All three of these conditions are met within the game industry.

Financial Power

Computer games resemble movies in more ways than being a form of entertainment. In both industries an average product takes between 18 and 24 months. Each requires a focused team with specific expertise to come together at different times during production—encouraging the use of contract and consulting workers. Each requires the investment of several million dollars in the creation of the initial product, but copies of the final product can be made for a nominal cost. Each relies on an electronic device to run them in the home. Each sells in the marketplace for $20 to $50. Customers are willing to purchase a number of these products during a year. Finally, in both industries it is difficult to determine which ideas will be popular successes and which will be enormous failures.

When a game is successful, there is no limit to the number that can be produced and delivered to the customer and there is no limit on how long the product can continue to be produced and sold. A successful product can generate huge streams of revenue long after it was first introduced into the market.

Kim & Mauborgne (1999) points out that customers purchase products based on the value curve that they perceive. When people choose to play computer games over watching television or participating in sports, it is because these games offer more value at a lower cost. Figure 16.6 attempts to compare the value curves of television and computer games to illustrate the package of benefits that are driving growth in the game industry. This growth has created a $10 billion industry that gets bigger every year and that is expanding into new forms of revenue generation to include in-game advertising, merchandising, competitions, and movie themes. Together these are generating profits that allow this industry to continually exceed its previous products.

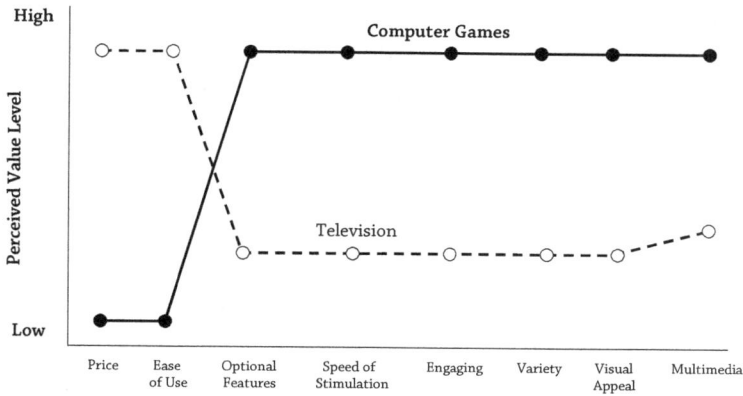

Figure 16.6. Perceived value of television versus computer games in a number of entertainment factors

Social Power

"This web of relationships between players—competitive, cooperative, and collegial—sustains the computer game industry, no less than the latest 3-D engine, facial animation algorithm, or high-speed graphics card."

(Herz & Macedonia, 2001)

21st century computer games have become social gathering points that have created a host of counter-cultures within society. However, these are unique from many previous phenomena, in that the social cultures within these games span the entire Internet-connected world. The popularity of World of Warcraft and Everquest is not an American phenomena, it has attracted a global audience. A group of characters on a quest in either of these games are as likely to come from different countries as they are to come from different states. The only force that still tends to bring together people from the same continent is time. Since the majority of players must balance a job with their game playing, this means that their schedules for entering the game often lineup based on the time-zone in which they live.

To support these social groups a number of open source products have been born which either competes with commercial products like Warcraft or that supplement the experience. Supplemental tools and web sites include newspapers, game asset managers, player tracking tools, team tracking tools, blogs on game events, and trading sites at which players can purchase game items using real money (Hof, 2006).

SourceForge is the dominant repository and collaboration site for open source software development, distribution, and coordination. In that repository, over 25% of all of the applications deposited and openly developed are computer games and supporting tools (Scacchi, 2004). This indicates that people are very eager to invest their time, intelligence, and emotions into the development of game software for which they will receive no financial return. They are eager to express their creativity, develop a reputation, or create a social group interested in their project.

As described earlier, "The effects associated with a customer network are not only a function of network size, but also network strength" (Shankar & Bayus, 2002). Computer game developers can overcome their rivals by creating stronger relationships between the players in their games. The strengths of the relationships in the network make the venture enduring and their enthusiasm for the experience attracts additional players and third party game developers. This personal attraction is one measure of the strength of the social network associated with specific game titles (Scacchi, 2004).

MULTI-INDUSTRY IMPACTS

"Having one foot outside your world means you can be less beholden to the people, ideas, and objects that would otherwise bind, and blind, you. Bridging multiple worlds, in essence, makes you less susceptible to the pressures of conforming in any one because you have somewhere else to go"

(Hardagon, 2003).

Game technologies are making this very transition into other industries. After establishing themselves firmly in the entertainment field, many game development companies have begun offering their products and services to a broad set of industries. When these applications are considered "serious business", like defense, medicine, city planning, and architecture, these tools are often referred to as "serious games". At least two recent books have been published under this title and describe a wide variety of applications of games to other industries (Michael & Chen, 2006; Bergeron, 2006). Table 16.2 summarizes the observations of several authors who have investigated this industry crossover.

Table 16.2. List of industries impacted by games and game technology.

Industry	Game Technology Impact
Military	Training soldiers and leaders in the tactics and strategies of war. Three dimensional modeling of equipment to illustrate or explore its capabilities.
Government	Ethics training for NASA. Project management training for the State of California.
Education	Augmenting classroom instruction in nearly every subject— English, math, physics, history, etc.
Emergency Management	Training emergency responders, firefighters, FEMA agents, and others to deal with disasters.
Architecture	Visually promoting major hotel, casino, and office spaces to potential clients.
City & Civil Planning	Lay out and experimentation with public services for a population of constituents.
Corporate Training	Orienting people to company products, facilities, and policies. Pilot and safety training.
Health Care	Educating patients on treatments, rehabilitation, and managing anxieties. The next generation of workout videos.
Politics	Presenting political issues and consequences of political decisions. Promoting candidates.
Religion	Interactive versions of sacred texts. Tools to teach religious history.
Movies & Television	Alternative form of storytelling known as "machinima". Tools for creating animation and 3D worlds.
Scientific Visualization & Analysis	Rapid display of objects under experimentation and physical forces acting on them. 3D display of data collected and analyzed.
Sports	Recreate live sporting events for review and for prediction of potential outcomes. Rehearse for critical "one time" events like Olympic ceremonies. Fantasy sports leagues in 3D.
Exploration	Prepare missions for NASA Mars Lander. Recreate environments around deep sea probes.
Law	Illustrate crime scene activities for judge and jury. Analyze crime scene data.

Sources: Michael & Chen, 2006; Bergeron, 2006; Casti, 1997; Maier & Grobler, 2001

All of these activities have been carried out in a different form without the use of game technologies. In most cases, those alternatives did not include a three-dimensional, visual presentation. However, in many cases, the human mind and imagination are not sufficient to visualize what is happening. Jay Forester, the father of System Dynamics, stated that, *"The mental model is fuzzy. It is incomplete. It is imprecisely stated. Furthermore, within one individual, a mental model changes with time and even during the flow of a single conversation. The human mind assembles a few relationships to fit the context of the discussion. As the subject shifts, so does the model ... Each participant in a conversation employs a different mental model to interpret the subject. Fundamental assumptions differ but are never brought into the open"* (Schrange, 2000). This lack of clarity and uniformity presents a valuable opportunity to apply game technologies, particularly 3D visualization and models of dynamic events.

Casti (1997) described the use of a football computer game to study possible final scores for Super Bowl XXIX between the San Francisco 49ers and the San Diego Chargers in 1995. Even though the computer game generated significantly different results from the real football game, this prominent scientist from the Santa Fe Institute, still considered the game to be a useful tool in understanding the dynamics of the Super Bowl performance.

Game companies and their technology did not move simultaneously into all of the industries listed in Table 16.2. Each industry requires a different level of sophistication for its applications. Therefore, game technologies, which are largely hosted on commercial computer hardware and software, had to wait for that foundation to grow powerful enough to tackle the problems in a new industry. Figure 16.7 provides a conceptual representation of game technologies moving "up market" into more demanding industries. The shorter shaded rectangles represent a fraction of all applications in that industry that could be impacted by game technology.

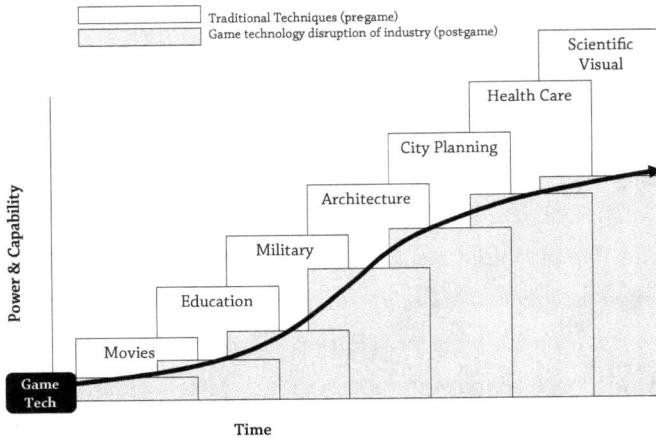

Figure 16.7. Game technology encroachment into other industries

Over time we believe the technologies being delivered by games will fracture into specialized applications and the common reference and root in the game industry will be lost in history. However, the current explosion in both computer and game technologies has allowed them to move so swiftly into so many different industries, that it appears that games are taking over all industries. This is directly parallel to the impact that the Internet had on industries of all kinds in the late 1990's. At that time it seemed that every industry was being turned inside out to become "Internet enabled" or to reinvent itself as an "Internet only" version of its previous identity (Porter, 2001).

Game technology impact is simply a concentrated, short-term transformation of the basic tools that have been under development in multiple industries for decades. Companies have been presented with new technology to improve their competitive position for several decades—small electronics, large computers, telecommunications, the Internet, the World Wide Web, personal computers, and now game technologies. Each of these offers some advantages and each company must determine whether to adopt the new technology. Those that choose the correct technologies will reap the advantages that they confer, and those that do not must continue to compete with the last generation of tools and technologies that they did adopt.

Figure 16.8 illustrates the adoption of repeated waves of technological advances. Companies that carefully adopt these can continue to improve their productivity, competitiveness, and reputation with customers. The dark shaded, multi-step bar demonstrates the path of a company that cautiously adopts these technologies. A company can benefit significantly by waiting until the new technology shows promise of displacing the old technology. In fact some authors recommend that companies not adopt new technologies immediately because that is when equipment costs and the level of uncertainty are the highest (Markides, 2005). Assuming that there is always time to catch-up to those who choose first and choose correctly, adopting more slowly can lower the risk of failure, lower the cost of entry, and increase profits. To enable this, a company may invest just enough in a number of new technologies to put itself in a position to leverage them into its products and services if necessary. This is a way to take an option on the technology without bearing the full cost and full risk of adopting it too early (Leonard, 1995).

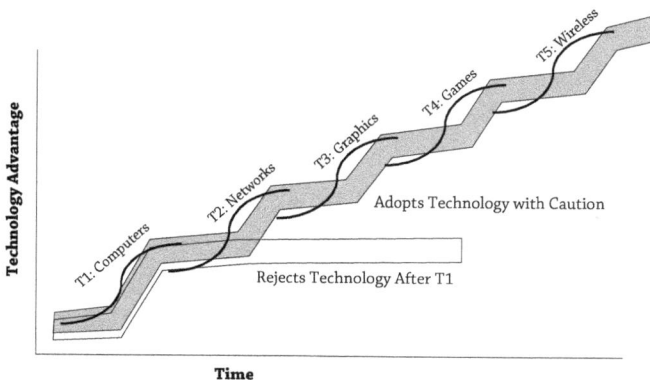

Figure 16.8. Conservative adoption of successive waves of technology

However, at the lower end is the lightly shaded bar in Figure 16.8. This illustrates a company that adopts just one or two waves of technology and concludes that anything after that is not important or valuable. When the ignored technologies do prove to be valuable, this conservative company will find itself competing at a significant disadvantage and potentially unable to survive.

GAME IMPACT THEORY

The very nature of games in Western society makes them a disruptive force. As Parker Brothers discovered in the late 19th and early 20th century, games have the power to influence society, but they must fit within societal norms. Today we see computer games extending their influence into the serious business of military operations, medical education, and emergency management training. In doing this, game technology is jumping the gap between entertainment and work. Throughout the evolution of electronic and computer games, this gap has kept this technology out of business, largely because games were not seen as "serious" tools. Games have been viewed as toys, not at tools for productivity. But the incredible power of the personal computer, graphics cards, broadband Internet connections, intelligent software agents, accurate physics models, and accessible user interface are making it impossible to ignore the potential of these "toys" to be applied to some very difficult problems in the "real business world".

Games as a Disruptive Technology

Once the barrier between entertainment and work was bridged, game technologies flooded into all of the areas listed in Table 2. As a relatively mature technology, games entered with a huge disruptive potential to the established players in those fields. Christensen's analysis of the disruptive effects of hydraulics on the steam shovel industry, mini-mills on large steel foundries, and small disk drives on their larger predecessors is a direct corollary to what is happening with game technologies (Christensen 1992 and 1997). These technologies offer significant computer and software power at a much lower price point than the solutions that are used in many industries (Figure 16.9). Games and serious industries were kept separated by the social stigma that has defined games as toys. This allowed the technology to mature significantly while that stigma dissipated. When it was finally gone, game technologies offered significant power for industry application and have been impacting these industries relatively rapidly.

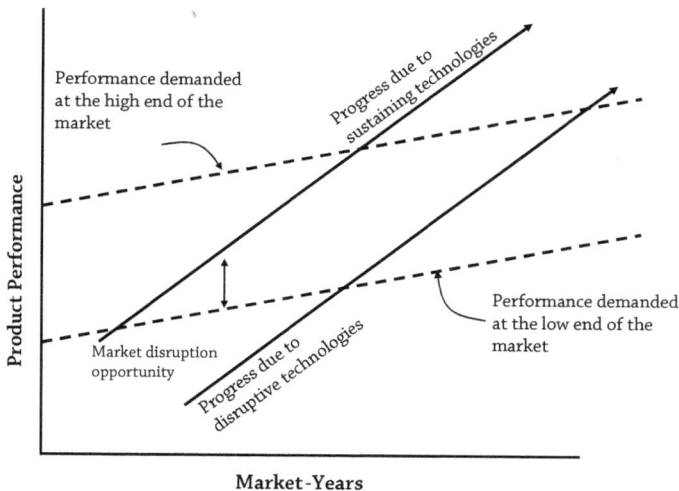

Figure 16.9. Christensen's theory of disruptive innovations explains how new technologies overthrow established businesses by offering better performance at lower prices. *Source: Christensen, 1997.*

Each industry that is assailed by these technologies faces its own set of arguments over whether games can perform serious work. But, those who insist that it is a passing fad are being bypassed by others who experiment with the technology and find a valuable use for it. As we showed in Figure 16.8, game technologies appear to be a natural next step from the graphics hardware and software that have most recently been adopted by military, medical, architectural, and other "serious" industries.

The power of the 3D graphics, accessible user interfaces, collaborative network connections, and intelligent agents is a persuasive argument. But, lower cost computer hardware and software to apply these technologies is making this technology irresistible and undeniable. In many cases, game applications run on machines that are an order of magnitude less expensive than their predecessors. Rather than paying $20,000 to $50,000 for specialized computer workstations, they can run on a $2,000 to $5,000 personal computer. Morris & Ferguson (1993) have pointed out that low-cost systems always swallow high-cost systems when this type of confrontation occurs.

The military has been one of the first and most avid adopters of game technologies. These games originated from military roots in the 1990's and contain many similarities with the training devices that are used to train soldiers. Therefore, the transition back into serious military applications has been much more direct than in other industries. Figure 16.10 extends Christensen's traditional graph of disruptive technologies to illustrate the multiple waves of game technologies that are transforming military simulation and training (Smith, 2006).

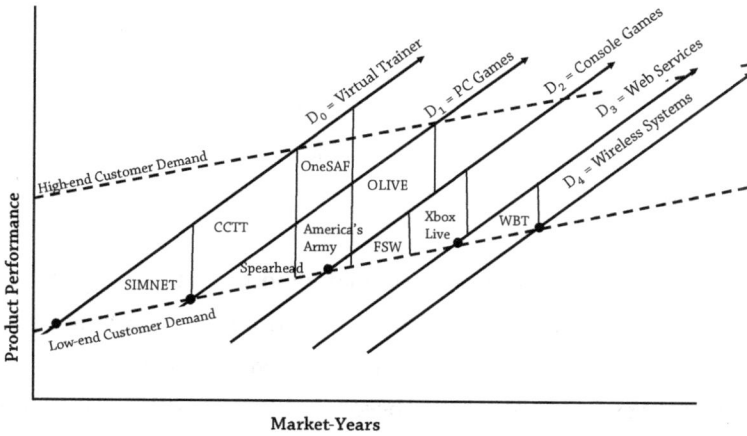

Figure 16.10. Multiple waves of game technologies that have already or are poised to disrupt the military simulation industry.
Source: Smith, 2006.

The first disruptive wave in Figure 16.10, labeled "Virtual Trainer", represents the creation of immersive simulators with three-dimensional graphics in the 1990s. Simulator Networking (SIMNET) and the Close Combat Tactical Trainer (CCTT) replaced a previous generation of devices by providing 3D computer generated worlds and networks to connect multiple training devices into the same world. They capitalized on the early Gould, Harris, and Silicon Graphics computers that brought 3D graphics to the engineering world (Miller and Thorpe, 1995).

The second disruptive wave labeled "PC Games" describes the emergence of SIMNET-like environments on desktop computers. The first set of applications like the game Spearhead demonstrated that PCs were capable of doing this type of work and encouraged other companies and government organizations to investigate new applications (Zyda, 2003; Lenoir, 2003; Mayo, 2005).

The third disruptive wave labeled "Console Games" describes the entrance of game consoles into the military market. These consoles offer yet another order of magnitude of reduction in computer hardware costs, dropping from a range of $2,000-$5,000 to $200-$500. This wave is just beginning in the military and it is not clear whether it will be able to overcome the licensing issues associated with developing a console game for a non-consumer audience.

The forth and fifth waves are speculative in that they suggest that technological advances will make it possible to run military training using game technologies through web-based services and wireless connections and that desktop hardware specifications will become a less important part of deploying these systems. Smith (2006) suggests that the pattern shown by the military adoption of game technologies will be repeated in other industries and that those industries should begin studying this issue themselves.

We suggest that game technologies will continue to move from one industry to the next based on five core forces of the technology and the environment in which it is emerging:

- Cost advantage of hardware platforms,
- Sophistication of software applications,
- Social acceptance of game tools,
- Successes in other industries, and
- Innovative experiments in the adopting industry.

Five Forces Driving Adoption

The five forces that govern the impact of game technologies on serious industries describe the attractive forces of these technologies into new areas (Figure 16.11). Where Porter's Five Forces model lists the competitive forces faced by an industry (Porter 1995), the Game Impact model represents the five compelling forces behind game technology adoption.

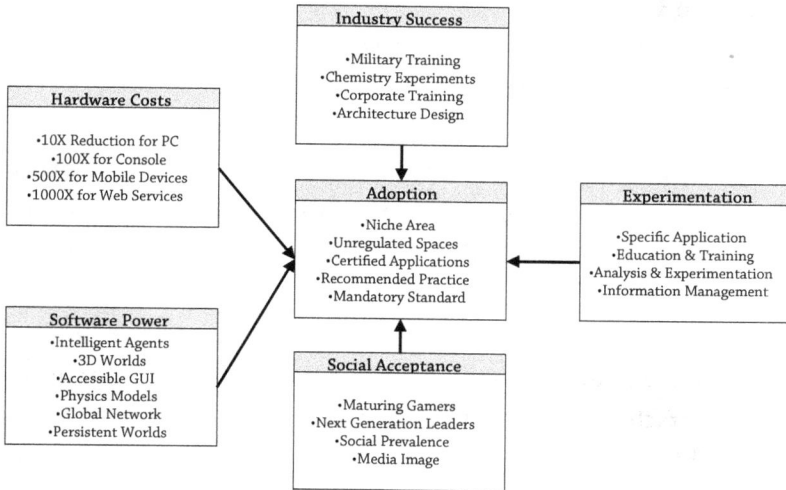

Figure 16.11. Game Impact Theory: Five forces behind the adoption of game technologies by diverse industries

Cost advantage of hardware platforms

Computer games are designed to take advantage of all of the power available on a consumer-grade computer. Their focus is on reaching the most customers based on the hardware that these customers have available. Therefore, unlike serious industries, game companies do not want to create a product that requires a new hardware purchase. As a result, these technologies are designed to be as efficient as possible, maximizing the amount of work that can be done on a consumer-grade computer. These machines are often an order of magnitude less expensive than a professional workstation, dropping hardware costs from the $20,000 to $50,000 range, down into the $2,000 to $5,000 range.

For games that run on the console platform, the hardware costs can drop another order of magnitude into the $200 to $500 range. These hardware savings can be significant for a company that must deploy its "serious applications" to hundreds of employees or customers.

Software Power

Game technologies are conquering some core problems that are shared across a number of industries. The ability to create a user interface that an average employee or customer can understand and operate is critical to a product's success. For a computer game, the goal is usually for the customer to understand how to use the product without ever reading a manual. Any instruction that is required is built into the game itself, allowing the customer to learn while they are using it.

Games also require clever and adaptive artificial intelligence to create game controlled characters that interact with humans in a realistic and engaging manner. Sophisticated AI has always required significant hardware resources and significant expertise to configure and run the system. Games fit this power into a consumer PC and provide scripting languages that allow a customer to change the behavior of the system.

Similarly, the 3D engine, physical models, global networking, and persistent worlds provide power that is impossible to achieve through any competing software products.

Social Acceptance

Games have largely overcome the stigma that they are just toys focused on play. The technology has persuaded most critics that these systems can be applied to serious industries. As the children who were raised with these games have become the leaders inside of companies and government organizations, the level of acceptance has increased significantly.

All of society has become accustomed to seeing 3D representations in courtrooms, medical facilities, museums, building designs, and military systems. After experiencing the advantages of this type of interface, people are much more willing to accept these technologies in serious products and services.

Other Industry Success

The television industry and the military have been two of the first adopters of game technologies. Television shows like Modern Marvels, Nova, National Geographic, and those on the Discovery and History channels have applied 3D visualization and physical modeling to illustrate the behaviors of animals, machinery, and the universe. The clear communications that these game technologies enable motivates other industries to experiment with them as well.

The military has incorporated many of these technologies into its training systems. Training devices for tank crews and company commanders all incorporate the 3D engine, GUI, physical models, AI, and global networking of games. The successes of these lead-users encourage other industries to explore them seriously as well.

Innovative Internal Experiments

As managers, programmers, and artists experiment with game technologies within an industry that is facing adoption, they discover useful methods for studying chemical reactions, understanding the stresses that occur between an aircraft and the atmosphere, evaluating the visual appeal of architectural designs, or delivering city services in a growing suburb.

When these internal experiments succeed in creating a new product or service, the established projects begin to experiment with the technology and look for ways to improve on their established practices.

Adoption Pattern

At the center of this model is the adoption pattern of the technologies. The adoption of game technologies in many industries may follow a pattern that is similar to that experienced by the military. It will begin in a niche area that is closely aligned with at least one powerful game technology. If successful there, it will be adopted for applications and activities that are not regulated. These are spaces where local groups define their own processes and measures of success. From this position, support will grow for the technology in a number of organizations and geographic areas. This will lead to some form of certified status of game technologies as an acceptable solution to specific problems. Success at this level will lead to it becoming a recommended practice in which the recognized regulating bodies will include it among the proven and preferred approaches to solving a problem. Finally, game technology may become a mandatory standard method of solving problems across the industry (Figure 16.12).

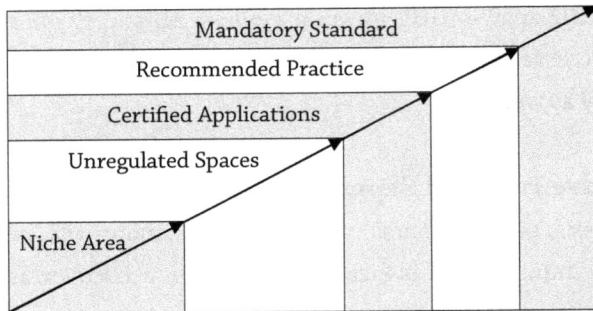

Figure 16.12. Potential stages of industry adoption of game technologies

The visual, auditory, and mental stimulation that come with games are often strong motivators for adopting and promoting the technology. Along with the flexibility that is built into the tools by core developers, these come together to create a very energetic lead-user community that contributes advances to the technology. von Hippel described this enthusiasm in the open source software development community (2001), and these forces appear to be even stronger in the game communities.

SEED OUT

Being relatively specialized, the game development community is highly interconnected. Even more so among those companies that develop applications for the consoles—e.g. Sony Playstation, Microsoft Xbox, and Nintendo Gamecube. Many small companies get their start by developing exclusively for a single platform, while larger companies develop for one or more platforms (Venkatraman & Lee, 2004). This has created a highly networked community with strong internal ties and that is somewhat insular. Figure 16.13 illustrates the tight relationships between console and game development companies. The companies that move game technologies beyond the entertainment industry are often outsiders who must fend for themselves in creating a new market. In fact, the companies that create the hardware consoles have shown little interest in and even active resistance to using the game console for non-entertainment applications.

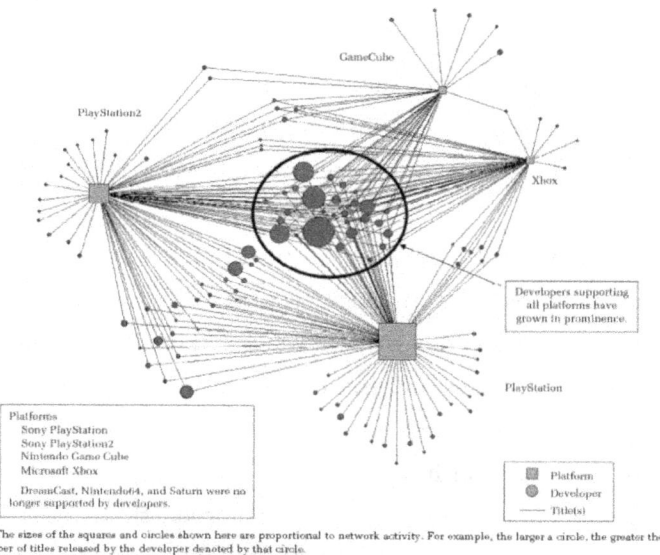

GameCube

PlayStation2

Xbox

Developers supporting
all platforms have
grown in prominence.

PlayStation

Platforms
Sony PlayStation
Sony PlayStation2
Nintendo Game Cube
Microsoft Xbox
DreamCast, Nintendo64, and Saturn were no
longer supported by developers.

Platform
Developer
Title(s)

ᵃ The sizes of the squares and circles shown here are proportional to network activity. For example, the larger a circle, the greater the number of titles released by the developer denoted by that circle.

Figure 16.13. Topology of business relationships between independent game developers and the manufacturers of the leading game consoles in 2002.
Source: Venkatraman & Lee, 2004

This social and business environment is generating splinter companies to serve the military, medical, architecture, and other similar serious customers. Over time, these splinters will become a unique community in their own space and will not be strongly aligned with their gaming origins. This type of relationship previously developed between the early game industry and its military technology parent, and will likely repeat itself as the game industry becomes the technology parent to "serious game" industries.

Determining the strategy for being successful and even dominant in these new splinter businesses will be important to these companies. Observing the key strategy that put Microsoft in control of the PC operating system, Morris and Ferguson have concluded that, *"A new paradigm is required to explain patterns of competitive success and failure in information technology. Simply stated, competitive success flows to the company that manages to establish proprietary architectural control over a broad, fast-moving, competitive space"* (Morris & Ferguson, 1993).

"Microsoft's insight was to realize that it was in an architectural contest and to take the appropriate steps, including steadily expanding the generality and scope of its systems to come out the winner" (Morris & Ferguson, 1993).

These ideas align with Utterback's model of innovation dynamics (Figure 16.14). In that model he points to the importance of the emergence of a "dominant design". This is a transition point at which innovations in product decline significantly as all companies converge on a standard design created by one of their members. From that point forward, the owner of the dominant design holds a controlling position in the industry. Following this inflection point, companies seek to innovate in the process and production spaces, looking for better ways to create or apply the dominant product design (Utterback, 1996).

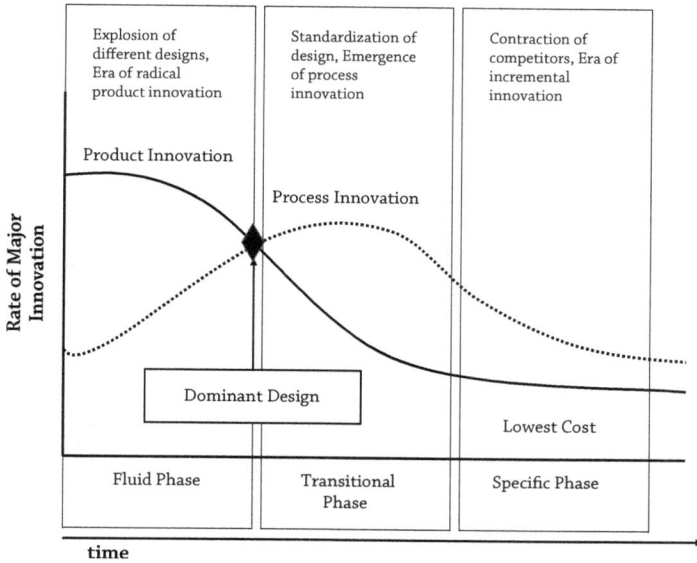

Figure 16.14. Utterback's model of innovation dynamics differentiates product, process, and cost innovation phases.
Source: Utterback, 1996

Using the five forces identified, we can speculate about other industries that will adopt these technologies in the near future. These might be:

- Television News,
- Sports Leagues (Real and Fantasy), and
- Space Exploration.

The first two can arguably be included in the entertainment category. However, game technology should be able to add more than an entertainment factor to each. Three dimensional animations are already transforming the way in which weather reports are given. Current weather maps and animations are generated by computer programs with similar roots to games. However, these do not go so far as to represent all of the buildings, roads, and social features of a city and to show how the weather will impact these in 3D. Progressive news stations may create a studio that shows the newscaster inserted into a 3D model of the city and its weather prediction. News reports on traffic conditions and road construction are amenable to these technologies as well.

Sports Teams and Leagues can use game technology to display, analyze, and predict the outcome of specific plays and games, just as Casti attempted to do for the Super Bowl. These may come to supplement the use of game films and provide the power to view the scene from any angle, as opposed to the fixed perspective of the film camera. Fantasy leagues can also apply this technology to go beyond the current scoring of fantasy team outcomes. Instead, a fantasy team could actually be created to see how they might have played had they been part of the same team.

Space exploration agencies and companies use 3D representations for design and experimentation with spacecraft. But, these tools are not derived from game technologies, but are industry specific and cost one or two orders of magnitude more than the gaming alternative. 3D worlds are also a natural control environment for missions on Mars. Since a radio signal from Earth takes several minutes to travel to Mars, it is not possible to "drive" an exploration vehicle with a joystick on Earth. But, driving in a local and accurate 3D model of the planet is an alternative way to create a travel path that can be sent to Mars for execution.

CONCLUSION

Game technologies have the power of technology, personal investment, financial profits, and social change behind them. In this paper we proposed a game impact theory that describes the forces that are driving the adoption of these technologies in a number of industries. The five forces described by this theory are:

- Cost advantage of hardware platforms,
- Sophistication of software applications,
- Social acceptance of game tools,
- Successes in other industries, and
- Innovative experiments in the adopting industry.

In addition to being technologically powerful, these tools and techniques are becoming more socially acceptable, even socially desirable, as the people who experienced games as children become the next generation of leaders in business, government, and the military.

"Why use simulations and games? An overly cynical answer to this question might be: because they get people enthusiastic and because we all have computers now!" (Lane, 1995). This cynical statement also captures some of the social/cultural forces that are driving this adoption. These technologies are overcoming the same types of resistance that confronted computers as tools for analysis and the Internet as a primary form of communication within business.

"The forces that hone games, and gamers, have more to do with anthropology than code" (Herz & Macedonia, 2001). As with the games introduced by George Parker over 100 years ago, these forms of entertainment test the edge of socially acceptable behavior and the use of one's time. They impact the social

relationships and cultural norms of a generation. The same can be said of business practices. It is the nurture of the individual that creates the current set of practices. As a generation of gamers enters the corner office and the oval office, these technologies will continue to gain acceptance. The five forces of game impact theory attempt to describe why this is happening and provide a framework within which managers and academics can evaluate game technology impacts on other industries.

Chapter 17

IMPACT ON POLITICS AND SOCIAL ORDER

T he intelligence community is eager to fully understand the impact of mobile technology, 3D gaming, and massively multiplayer gaming on politics, economics, and security. Analysts are particularly interested in answering the questions below. Though each question could serve as the basis for a study, the brief answers outline the opportunities and threats posed by these new technologies.

Virtual Worlds, Games, and MMOGs are creating a digital forum for learning, participating, exploring, and creating ideas. This forum has no physical limits and obscures physical, political, religious, and geographical differences. "On the Internet no one knows you're a dog" (Peter Steiner, *The New Yorker*). The same is true for virtual worlds. A collective intelligence has emerged and is making everyone in these spaces smarter and closer knit.

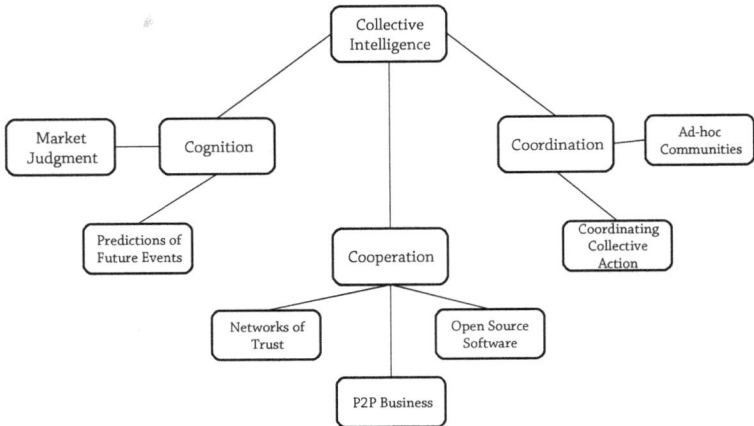

How is 3D modeling/simulation/gaming bridging the gap between the real and the virtual worlds (real + virtual = mixed reality)?

Environments like World of Warcraft, Second Life, Active Worlds, and others are demonstrating the feasibility of creating very large shared virtual spaces. Second Life has established an economic environment in which it may be advantageous for companies to have a presence. It is perhaps the first instantiation of a 3D World Wide Web. This will allow companies like Best Buy to create virtual stores that mimic the brand style of their physical stores. Hopefully, companies and organizations will use virtual spaces to overcome some of the limitations of physical spaces, not just duplicate them.

I have held discussions with members of the Department of State to explain the power of virtual spaces. They do not need to create exact copies of their real-world embassies. But can use these virtual spaces to present America in a much more positive light than do our current embassies. For example, the virtual spaces will not need physical walls, barbed wire fences, guards, dogs, metal detectors, and search procedures. These are the up-front real-world face of the embassy and not necessarily the message we want to carry to the people of these countries. In fact, the Second Life Embassy does not necessarily need walls, doors, and furniture. It may be more conducive to represent spaces with icons of America like the Statue of Liberty, Golden Gate Bridge, and Grand Canyon to convey the opportunity and beauty of the country. Virtual embassies become more like Walt Disney World's EPCOT as a showcase to the world.

Over time I expect that everything with a physical presence will have a virtual presence. One instantiation of this is an exact copy of the physical world in the virtual space. This is not the best solution. Having the Great Wall of China and the Washington DC Mall on opposite sides of the planet is not optimal. The virtual world may allow us to eliminate the gaps between people, organizations, and countries which have grown out of physical locations and long distances.

The virtual representation of the world could place everything within equal distances of everything else. It is a hyperspace in which distance is measured through many different dimensions and can be greatly minimized. The makers of Second Life may or may not realize the value of this. Currently they are dealing with a customer-base that is very physical world oriented and it will take a few years to abstract around this way of looking at "worlds".

What role will ubiquitous computing play in furthering the emergence of mixed reality games?

Game spaces and experiences will be extended to all electronic devices. Cellular phones and wireless computing devices (Gameboy DS, Palm, Pocket PC, Mylo, Nokia 770, etc.) will be linked into the game world. This is happening first in Japan, Korea, and China. They seem to have a greater acceptance for small devices and small representations of the world. In the US we prefer bigger displays and bigger feelings for the world—perhaps an extension of our Wild West perspective of the world.

There are already services which allow you to track the physical locations of friends in your social network. As 3D worlds like Second Life expand, they will begin to integrate these types of services from social networks. Future social networks will allow people to share information, track each other, and integrate this information with a virtual representation of the real world. This is a perfect tool for coordinating emergency response teams, SOF missions, and terrorist actions.

How can MMOs (Massive Multi-player Online Games) like Second Life and Project Entropia positively and/or negatively impact politics and democracy? Will it have any enduring effects at all? Will the effect be different in closed vs. open societies?

If Second Life, or a successor, becomes the World Wide Web in 3D, then it will have an impact similar to the first WWW—huge! Internet/WWW locations are currently independent and separated from locations in physical space. This has been fantastic for eliminating barriers. But it has also limited its use as a collaboration, planning, organization, and execution environment. The Internet and WWW have taught everyone to think and act independently from physical presence. They will be able to carry this mindset with them into the 3D virtual world and act with or without reference to physical location. Everyone is familiar with teleportation in 3D worlds. This eliminates the inconveniences of distance. But it also allows them to reenter the 3D world at a chosen point where an important event is occurring or being coordinated.

Asking whether MMOGs and similar spaces will be enduring is like asking whether the WWW will really have an impact on the Internet.

Positive Impacts:

These 3D spaces are being created by people in democratic countries. The values they hold are becoming the values of the virtual world. Therefore, the environment should encourage greater democracy in its participants. I suspect that the developers are more democratic than the US political body. They will probably allow/encourage more democracy than even the US is comfortable with.

Negative Impacts:

Those who oppose democracy can use the 3D environments to create dioramas that demonstrate events in which democratic approaches are weak and lead to suboptimal results. Like everything else in 3D, these should be able to carry the message to a much larger audience than does a textual document (e.g. Television vs. Newspaper in carrying a message).

How will civilian or criminal groups use ubiquitous computing to harm state security or other citizens' privacy or property?

These allow collaboration in 3D for criminal or terrorist activities. They create a rehearsal environment that maps physical location, but which is potentially hidden, private, and undetectable from the outside. Wireless devices could be very useful to coordinate the actions of many people through all phases of an attack.

Privacy is becoming very tenuous now that the fractional cost of getting information is falling to zero. A threat organization could create their own Google engine (or leverage the real Google) to search the entire Internet for information on places, people, and organizations of interest. This data could then be fed into an analysis engine to identify what is useful, extract it, organize it, and apply it to their ends. Imagine any organization being able to rapidly collect the names, addresses, family members, telephone numbers, day care centers, schools, and restaurants associated with every member of Congress or some similarly significant group of people.

With information and coordination tools, it may be possible for a threat organization to successfully pull off dozens or hundreds of simultaneous attacks that stress the abilities of defense organizations to respond. Would it be possible to start the wheels in motion to attack all US Embassies simultaneously? Could this be coordinated to completion without being detected?

We know that terrorist groups already use video games as training tools and mobile technology to execute attacks. Will the convergence of mobile, 3D modeling/simulation, and ubiquitous computing technologies improve the ability of dark mobs to plan ("previsualize"), coordinate, and carry out violent attacks?

Mobile computing devices allow people to be in the physical space and in the virtual space simultaneously no matter where they are in the physical world. This means that real-time coordination in 3D spaces is very possible.

The 3D world may also become a communications medium in which the results of the attack are captured, displayed, and replayed for all to see. There are some interesting new devices like gamecaster (a movie camera for in-game filming) that should make in-game recordings of much higher quality. These may lead to the creation of Machinima that rivals the creative and technical quality of Hollywood movies. Coupled with virtual worlds that match the physical world, it would be possible rehearse, film, analyze, and improve actions in the virtual world with details that are very close to their real world counterparts.

Chapter 18

THE LONG TAIL IN MILITARY SIMULATIONS AND GAMES

T he commercial market for computer entertainment products has financed and driven the creation of a very robust and growing industry in computer games. Commercial demand for these products has been so strong that it has driven the design of computers, CPUs, graphics cards, sound cards, and a number of other devices. The money earned by these game companies and their supporting hardware providers has allowed them to invest in research in both software and hardware that has moved the entertainment industry to the head of the technology curve. Today, game companies and commercial graphic chip developers lead technology development, a position that was held by the government, military, medical, and heavy industries throughout the 20th century.

Commercial demand has allowed complex software and powerful hardware to be delivered to customers for under $100. During the 1990's, leading technologies in these areas sold for hundreds of thousands of dollars. Coupling the immense power of these products with significantly lower costs has made it possible to deliver games and simulations to more customers than could be reached at 20th century prices. Game technologies have the power to enable the emergence of a long tail in military simulation that is very similar to that which already exists in digital music. Like the introduction of the Xerox copier, the availability of the technology creates its own demand. *"The power of the Xerox copier did not lie in its capability to replace carbon paper and other existing copying technologies, but in its ability to perform services beyond the reach of these technologies. The 914 [copier] created a market for convenience copies that had previously not existed"* (Hammer & Champy, 1993).

THE LONG TAIL OF DIGITAL PRODUCTS

Chris Anderson's 2004 article in *Wired* magazine introduced the idea of a "long tail" in digital product sales. He describes how the Internet has lowered product delivery and stocking costs so that it is now very profitable to sell hundreds of thousands of unique products rather than just the "Top 50" products found in most retail stores. The idea is most powerfully applied to music, movies, games, and similar digital products.

Traditional products like computers, textiles, and music CD's all result in a physical item that has both fixed and variable costs associated with its creation and delivery to a customer. Fixed costs refer to the land, facilities, machinery, and similar items required to bring a manufacturing capability into existence. These costs present a large barrier to entry for all companies that wish to compete in an industry. This barrier is an impediment to newcomers, but a protection to established firms. It provides some insurance that everyone will not build a competing facility on a whim and try to enter a market. Variable costs are associated with the materials, labor, packaging, shipping, and storage that are associated with each item that is manufactured. For a textile item like a dress shirt, these include the cost of fabric, buttons, and thread (materials); cutters, sewers, and folders (labor); pins, stays, and bags (packaging); semi-truck delivery (shipping); warehouse space, electricity, and HVAC (storage). Companies work hard to find ways to reduce both their fixed and their variable costs. Lower costs allow them to sell at more competitive prices and to generate higher profit margins.

In this digital age, we have discovered that some products are perfect for changing this cost model and the relationship with the customer. A recorded song or movie is primarily a digital product. But recording the product (song, movie, or game) onto a CD-ROM and shipping it to a store for sale turns this digital product into a physical product, with all of the fixed and variable costs associated with that. If it is possible to deliver the digital product to the consumer without ever turning it into a physical product, that would significantly reduce the variable costs associated with each song, movie, or game.

Companies like iTunes, Rhapsody, and Napster have done this for songs. YouTube, Yahoo! Video, and Lodgenet have done it for movies. Valve Software and Real Media have done it for computer games. When they eliminate the physical costs they become much more competitive than traditional vendors who continue to deliver these products as physical items. Digital product companies have also discovered that in any given fiscal quarter, at least one customer purchases every single song or movies loaded in the database. That is enough to pay for the storage cost of the least popular title and to make a positive contribution to reported quarterly earnings.

MILITARY SIMULATION IMPLICATIONS

Current military training simulations have the cost model of a physical product. Helicopter flight simulators, for example, include a significant physical hardware suite that must be produced, shipped, stored, and installed at the customer site. Even constructive wargames are usually tied to a unique suite of computer equipment that must be procured and delivered to the customer. These simulations also require that system experts accompany the hardware and software delivery to insure that it is properly installed and configured. For these systems, there does not appear to be an opportunity to take advantage of the Long Tail effect in digital products. These represent the "Short Tailed" limitations of demand for physical products.

However, the military is increasingly using commercial computer games and the technologies associated with those as a foundation for lighter, desktop training systems. When these simulations can be limited to a software product that will run on a customer's existing computer hardware, they have taken the first step toward being "long tail enabled." If the products can further be designed so that an expert does not have to accompany them to insure that they are installed properly and taught to the customer, then they have taken the second necessary step toward being a "long tailed" product.

Any training simulation that can be delivered to a customer's existing computer and used without the help of an expert is a long tail product. Such products do not have to have their roots in computer games. The computer game is one example in which this transformation is becoming possible for military simulations. Games are a commercial business area that has been challenged to eliminate special hardware and expert human support, and which has accomplished this successfully.

Some of the significant advantages of all-digital simulation products are:

- Delivery to customers anywhere on the Internet,
- Accessible far beyond the reach of physical experts,
- Distribution via viral marketing forces,
- Significantly lower variable product costs, and
- Potential to serve much smaller customer niches.

DIGITAL TRAINING SYSTEMS FOR ALL JOB DESCRIPTIONS

The focus of military training simulations has historically been on devices that prepare soldiers to perform life-threatening operations. The simulation device is one tool that allows soldiers to develop their expertise with equipment, in teams, or as commanders, without risking their own lives or those of others. Historically, these devices have been custom built to teach specific lessons and skills, which has typically led to a combination hardware/software or physical/digital solution. These custom solutions can be very expensive to create, with initial development budgets of more than $100 million and per suite prices of $1 million each.

At these prices, the military can only afford to address the most lethal combat operations before it has exhausted the financial and talent resources available in both government and industrial organizations. However, if these simulation systems could be based on commercial desktop software tools rather than custom developed products, it may be possible to create more systems in less time and at lower costs. Game engines and their associated development tools enable this approach to simulations. Therefore, as we are developing digital-only training systems, we should be able to do so for significantly less than is spent on current training simulators.

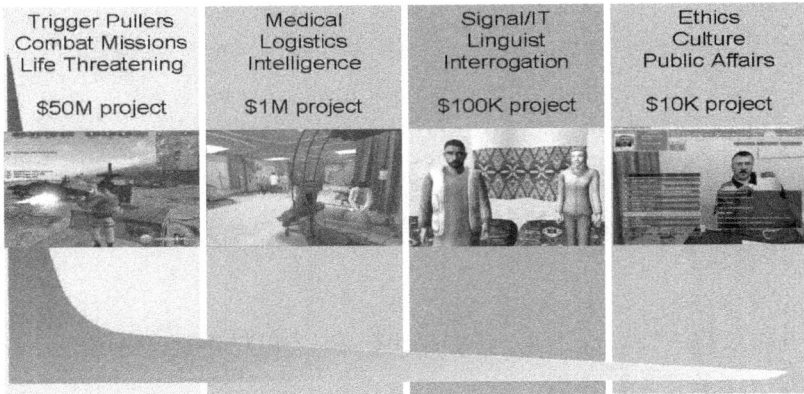

Trigger Pullers Combat Missions Life Threatening	Medical Logistics Intelligence	Signal/IT Linguist Interrogation	Ethics Culture Public Affairs
$50M project	$1M project	$100K project	$10K project

Commercial game technologies may enable the emergence of a "Long Tailed" product line with a number of lower cost simulations to serve specific training needs.

These tools should allow us to move down the tail of the demand curve for simulation and meet the needs of smaller niches that until now had to use other means for training. If "digital-only" simulations meet the needs of medical, logistics, intelligence, signal, linguist, and public affairs personnel, then we may well be entering a period in which we can afford to address this need.

Breadth of Army MOS

The U.S. Army identifies its internal job descriptions as Military Occupational Specialties (MOS's). An unofficial list of these compiled on the Wikipedia web site identifies 316 MOS's within the Army. These are grouped into 26 categories and divided into separate lists for enlisted and officer ranks. This list presents one map for determining how many game-based simulations could be built in this long tailed world. However, the list does not identify all of the tasks, operations, or missions that are a unique combination of multiple personnel skills and equipment. It is just one means for understanding the breadth of jobs and schools that exist in one military service. A cursory analysis of this list led to the initial conclusion that game-based simulations could be used for as many as 16 of the 26 major job categories—or about 60%. There are a number of jobs for which simulation is probably not a viable, necessary, or preferred means of training. Some of these are MOS 42R9D (French Horn Player), 79T (Recruiting and Retention Officer), and 92S (Shower/Laundry and Clothing Repair Specialist). However, there are a number of jobs on the list for which a simulation may be very beneficial, but to our knowledge is not yet available. These include MOS 15G (Aircraft Structural Repairer), 25W (Telecommunications Operations Chief), and 68D (Operating Room Specialist).

Accessibility

As described earlier, the keys to enabling the long tail in military simulation are (1) digital-only products, (2) globally downloadable, and (3) non-expert user friendly. Enabling these requires more than just the creation of a digital training system, it also requires the existence of a global information network that serves high-bandwidth, real-time data to soldiers.

Digital access to the simulation library is a good start, but that must be supported by a repository that is online 24/7, easily searchable, and that can verify the identity of the people accessing the data. The users of the simulations will also need tools that allow them to create or modify scenarios that teach the lessons they are attempting to learn.

Key Jobs, Key Trainers

The military has traditionally focused their training on jobs that involve lethal engagements. As game technologies open up the community of soldiers that can have simulation-based training, the metric of lethality will remain one that determines the priority for creating and releasing a new system. An analysis of the fatalities and injuries experienced in the current Iraqi War may aid in identifying the most important tasks for which additional training is needed.

The casualty data compiled at icasualties.org identifies and ranks the causes of death in this war (Cutler, 2007). At the top of the list is the Improvised Explosive Device. In fact, this cause is higher than all of the next six causes combined. Until recently there was no IED defense training device and there remains no computer-based training to prepare people to identify, avoid, and defeat IEDs. Desktop, game-based training tools are one means of teaching the skills and procedures for handling these situations.

Another source of fatalities that may be reduced through desktop simulations are non-hostile vehicle and helicopter accidents, which accounted for 323 fatalities between March 2003 and April 2007. There are a number of driving simulators and driving games that could be readily modified to teach specific skills that are leading to accidents in daily operations.

CONCLUSION

Computer games and the technologies that they have advanced offer the opportunity to deploy a game-based training simulation to every soldier in the world. They allow us to escape the burden of special hardware devices, centralized training facilities, and the staff of experts required to run existing systems. Though the existing systems will continue to be a central part of the military training regime, games could enable us to extend training beyond these and create a long tail of products that fit into unique niches. In some cases, game-based systems will also offer a lower cost alternative to existing systems. This may become more important when defense budgets diminish or funding is directed away from training systems to support operational equipment and missions.

The long tail of digital products described by Chris Anderson can potentially express itself in the military simulation industry. Desktop game technologies are one of the tools that can make this happen.

Chapter 19

Gaming the US Army

A pple Inc. CEO Steve Jobs is famous for his flashy, headline-making introductions of innovative new products. While the iPod, and more recently the iPhone, generated a great deal of publicity right from the outset, many innovations are hardly noticed at first.

In fact, often seemingly trivial developments can be harbingers of big change. When a small, regional airline began operating out of Dallas's Love Air Field, no one foresaw that Southwest Airlines would fundamentally alter air travel. And when a guy started selling Pez candy dispensers on an obscure website, few imagined the impact eBay would come to have on a variety of industries.

These stories of course are well-known to most everyone now, but few people are aware of an apparently insignificant event in 1995 that fits the disruptive pattern we've seen before. Working on a shoestring budget, a U.S. Marine Corps lieutenant and a sergeant had a radical idea: To try to alter the popular

"Doom" video game, in which players use a variety of weapons to fight electronic foes, for use as a military training tool.

The result was "Marine Doom," a (by today's standards) rather simple video game that could be used to teach soldiers certain skills at a low cost. The developers were hoping to find a way to boost training in an era of significant budget cuts and came up with a cheap, simple, and convenient—in other words disruptive—solution.

Yet, in spite of all of its benefits—and considerable media attention at the time—Marine Doom arrived too early when game technologies were too primitive for serious work like this. The story of why this first effort was unable to gain traction and how in the intervening years gaming has grown into a valuable training tool offers meaningful insights into how small disruptions can ultimately make a big impact.

A FAILED FIRST FORAY AND A SUCCESSFUL SECOND EFFORT

Much like the minicomputer industry's reaction to early personal computers, many in the military brass considered Marine Doom and other early training games distractions from the real work of developing and using traditional multi-million dollar simulation systems to train soldiers.

Although weapons-related games and soldiering would seem like a natural combination, the deployment of games as a serious military training tool has been anything but simple. These low-cost technologies were attempting to disrupt a decades-old, monolithic defense industry that was well entrenched in the purchase processes of the military.

Simply put, using consumer-level PCs and video game technology seemed all wrong when seen through the traditional ways that the Army trained its soldiers. Mid-level officers were used to conducting classroom training, where they would whiteboard scenarios and interact with soldiers. Senior leaders were accustomed to developing and purchasing state-of-the-art, multimillion-dollar simulators that took years to commission and build.

While the chain of command did not embrace gaming, it did not kill it entirely. Faced in 2000 with record low recruiting numbers, the Army returned to gaming to help attract new recruits. America's Army Game, an online video game developed internally that relied on the Unreal game engine created by Epic Games Inc., took center stage in the effort.

The game attempts to simulate the experience of an Army soldier by allowing users to play out a variety of scenarios. Instead of just containing fight scenes, the game tries to educate users about the Army and the various career paths different soldiers can take.

Unlike most war-based video games that emphasize killing enemies, America's Army awards points for factors such as teamwork, responsibility, and good values—traits the Army deems essential.

The goal of America's Army was to attract young men and women to their local recruiting offices. It was a hit. Enrollment went up, hundreds of thousands of people downloaded the game, and the effort became a public relations sensation for the Army.

Why did military leaders embrace America's Army but shun Marine Doom? For one, junior officers who had supported the idea of games-as-training-tools when Marine Doom came out had risen to more senior decision-making ranks and by 2000 were in a position to encourage and fund this new project.

More importantly, America's Army targeted a "foothold" market with much less rigorous standards for acceptance than had Marine Doom. Preparing soldiers for battle is a core function of the military: Any mistakes would have major consequences. This meant that new training tools or processes had to meet to extremely high internal standards.

America's Army, on the other hand, was just a marketing tool. If the product flopped, no lives would be lost and, while recruitment might not get a boost, it was unlikely to plummet either. As such, the project and the final product did not receive the same level of scrutiny from the Army's most demanding users. Like most successful disruptions, this allowed the project to launch and build quickly, identify elements that were successful, and retool as needed.

HAVING SECURED A FOOTHOLD, GAMING MOVES UPMARKET

The success of America's Army spawned a renewed interest in military gaming and the technology mushroomed into dozens of new training applications. A number of Army organizations began to invest in their own game-based tools, creating training systems for things like learning how to control robots, use new rifles, steer remote-control machine guns, and convey basic "Army 101" information.

The legendary Defense Advanced Research Projects Agency (DARPA), the technology research arm of the Department of Defense that developed the progenitor of the Internet, began working on gaming applications, in some cases in conjunction with private companies. DARPA focused on identifying ways to cheaply and easily adapt commercially available games to meet specific military training needs.

Today, the Army is continuously identifying new opportunities to expand the role of gaming technologies in simulation and training. The long-term goal is to work with smaller commercial gaming system manufacturers and commercial software developers to create new individualized training systems for all soldiers. To get there, the Army is redefining how it trains and educates—and thousands of existing processes and technologies could be disrupted.

One of the biggest successes is Ambush!, a computer simulation game now used in the field by the Army. Ambush! trains soldiers how to extricate themselves from the deadly confrontations that occur regularly in parts of Afghanistan. In fact, troops have embraced the game so fully that one of the main highways in Afghanistan has been nicknamed Ambush Alley.

As is the case with most successful disruptions, a number of outside factors contributed to the successful adoption of gaming by the military.

Ongoing improvement of the innovation: In the early years, available games were so simplistic that many people could not imagine using the technology in demanding military contexts. The development of the Unreal game engine and other technologies have allowed richer levels of detail and larger battlefield maps, making simulation games far more realistic and scaleable than the earlier Doom-based games.

Shifting context: The nature of the military challenges faced by the Army has changed from a World War scenario in which large forces of thousands of soldiers are mobilized across countries to more fragmented battles in which a few dozen soldiers fight enemies from one street corner to the next. Current gaming technology is better suited to simulating this environment than are the incumbent systems that created complex, large-scale battle scenarios.

Changing "customer" needs: To respond to current situations, soldiers require individualized training, whether it is teaching a medic how to evacuate a fallen soldier from a street, training an interpreter how to interact with local leaders, or teaching a convoy driver how to spot a potential ambush. Most traditional Army training has tended to emphasize widely used skills, not the customized learning these new technologies enable.

THREE LESSONS

The growth of low-cost gaming technologies in the Army offers a number of lessons about how successful disruptions can take hold and continue to grow.

Flexibility enables disruptive success

Relative to existing military combat simulators and training tools, video games are simple and low cost. At first, this made the offerings poorly suited to accomplishing the highly demanding and important job of preparing soldiers for battle.

But some in the military did see potential applications and exploited the flexible nature of the innovation to suit a different, less demanding niche: recruiting. Once gaming had significant success in one area, innovators were able to build on this to move into more demanding tiers.

The flexible nature of gaming technologies is altering industries well beyond the military. Linden Labs's Second Life, a self-described "online society within a 3D world, where users can explore, build, socialize, and participate in their own economy," began as an advanced social network where people could use avatars to interact and make virtual transactions online.

For its first couple of years, Second Life predominantly attracted individuals, especially teenagers who wanted a fun place to interact online and did not demand top-end functionality. Because the virtual environment was incredibly flexible, Second Life was able to dramatically improve over time, refining features users liked, squashing those they did not, and providing greater graphics and memory capability.

Now, demanding corporate marketing departments have found numerous uses for the newest version of Second Life. Starwood Hotels is premiering its newest property brand, named "aloft," in the online universe, including undertaking a virtual construction of a marquee building. Dell Computer sells actual PCs through the virtual store, while Toyota allows users to buy virtual versions of its Scion xB to users who want to drive their avatars around the Second Life universe. And finally, both rounding out and summing up the potential of gaming, IBM had a virtual meeting in Second Life to discuss the effects of multiplayer games can have on businesses.

Disruptive innovations come from disruptive suppliers

The Army struggled with this lesson because it had been well served by incumbent, large suppliers for many decades. In the past, these suppliers were able to anticipate the military's needs and to deliver cutting-edge, innovative products and solutions.

But, just as the Army failed to recognize the full value of games as training tools early on, traditional suppliers of military training and simulation systems failed to recognize the changing needs of the Army itself.

For example, one of the great advantages of new PC-based game systems is the high level of detail it offers, which is valuable in simulating street-to-street combat. But established projects, organizations, and investments in technology prevented existing contractors from creating these newer and higher fidelity tools.

Why? First off, existing providers were raking in handsome revenues from improving existing technologies and products, which the military gladly purchased. And, no one would expect a company that is organized to develop high-margin, complicated, costly simulation systems to suddenly prioritize making cheap, inferior solutions. To succeed in such a low-margin market, an incumbent would have to organize itself entirely differently—not an easy task.

In contrast, computer game developers had been cranking up their ability to deliver high fidelity at a low cost for years. These companies are naturally inclined to recognize that high levels of detail are crucial in new types of simulations—it's the same attribute of performance that young gamers in the consumer market have been demanding from video game makers for decades.

Although it may seem surprising that the entertainment industry should be a source of technological innovation for the Army, it is worth recognizing that computer games have become a major industry, with annual sales approaching $50 billion.

Innovative organizations evolve

When Marine Doom launched, the Army failed to see and build upon the (to some) obvious potential gaming held to improve and lower the cost of training. One reason for this was that senior leaders simply did not understand the potential of the new technology. They were accustomed to purchasing expensive, complex systems that had undergone rigorous analysis, not doctored versions of video games played by teenagers.

What could they have done differently? They could have tried to learn more from the people who were embracing the disruptive technology early on. Soldiers themselves were a great early indicator of the value of the innovation. They were naturally inclined to use games as a training tool because they had grown up playing such games.

The early success of Marine Doom was a signal of change, but this signal was not properly interpreted. It was seen as a distraction from the main business of training and simulation, not as an early prototype that could be embraced, tested, refined, and built upon.

Additionally, the military, like many large organizations, was hindered by an overly complex purchasing system and a reliance on traditional suppliers. Although the Army has tried over the years to court new suppliers, it has been unable to draw a lot of small, disruptive partners for one main reason: The Army is a complex customer to work for.

Smaller companies focused on game technology frequently concluded that the Army was either unwilling or unable to work with them. Decisions were made too slowly for the smaller companies' sales cycles and Army leaders were unprepared to make significant commitments.

Within this tension, there is clearly a role to be played by suppliers with a mastery of the disruptive game technologies and a culture of understanding and dealing with the acquisition processes of the Army.

Disruptive forces are unfolding all around America's armed forces. The nature of world threats has changed and the technologies for dealing with those are changing as well. The emergence of game technologies as alternatives to many of the established tools for training is just one of the disruptions that are forcing the Army to adapt to the needs of the 21st century.

Chapter 20

SENSOR PERFORMANCE IMPROVEMENTS

M ost game objects are interested in two questions, *"Who can I see?"* and *"Who can I shoot?"* When this question is asked by a game controlled object, usually referred to as an AI or NPC, it presents a problem of determining which objects are within range. Unfortunately, this can sometimes mean searching through the entire list of objects in the game and subjecting each to a line-of-sight (LOS) or range check. This is hugely inefficient and completely unnecessary.

In virtual worlds, the primary organizing characteristic for interactions between objects is geographic location. Objects tend to interact with those in close proximity—this includes sensor detection, weapon engagement, communication, and the exchange of supplies.

Game programmers must manage lists of dynamic (living, breathing, moving) objects so as to retain the geographic relationships between them.

The commonly used linked lists and many tree structures do not retain this information. A linked list usually manages all of the objects, but does so in a nearly generic manner so that the order of the list does not contain any useful information. Trees may embed some useful information in the data structure, including geographic location. However, implementations like quadtrees are generally applied to static terrain rather than dynamic objects. This gem describes a simple method for managing object lists using geographic grids as a means of significantly reducing the computational work required for range and line-of-sight decisions. Managing objects geographically can reduce the number of line-of-sight and other related geographic checks on a large battlefield from on the order of 1 million to a few dozen.

Some games enjoy a naturally occurring grid system for managing their objects. When the game level represents the inside of a dungeon, building, or space station, the world is naturally divided into rooms. A "portal engine" will take advantage of this to identify near objects according to the room where they are located. Visible or interactable objects exist only in the local room or an adjacent room with a portal leading to it. Though the game may actually contain several hundred objects, only two or three may be in the room with the object that is looking for targets. This significantly reduces the amount of work required to identify objects with which to interact.

Unfortunately, not all games are so nicely structured. Large open battlefields or spacefields can contain hundreds or even thousands of objects. When there are no walls, mountains, or forests to separate them, all objects effectively exist in the same room. This presents a difficult computational problem for an AI agent that is trying to identify the closest or most desirable targets. Svarovsky described this problem very briefly in the first volume of *Game Programming Gems* (Svarovsky, 2000). In this gem we explore it more deeply, explain why these open spaces are so difficult to deal with, and present a relatively simple solution that has been implemented in a number of simulations. Pritchard also explored tile-based LOS in the second volume of Game Programming Gems (Pritchard, 2001). But, he limited his discussion to detectability from the perspective of the human player, leaving the actual detection outcome to the graphic rendering engine and the human eye. This gem includes complete detection decisions for AI agents who rely on range and LOS calculations entirely in the game engine.

QUADTREES AND OCTREES

Graphics programmers are very familiar with quadtrees. These structures nicely divide a two-dimensional space into four smaller "child spaces". This process can be continued from generation to generation until the entire 2D space is divided into small squares that contain the terrain data (Figure 20.1). A sensor viewing frustum overlays a specific sub-set of these squares, identifying which grids need to be drawn. Therefore, it is not necessary for computer hardware to render the entire terrain database, when only a fraction of it is viewable by a sensor (Ferraris, 2001).

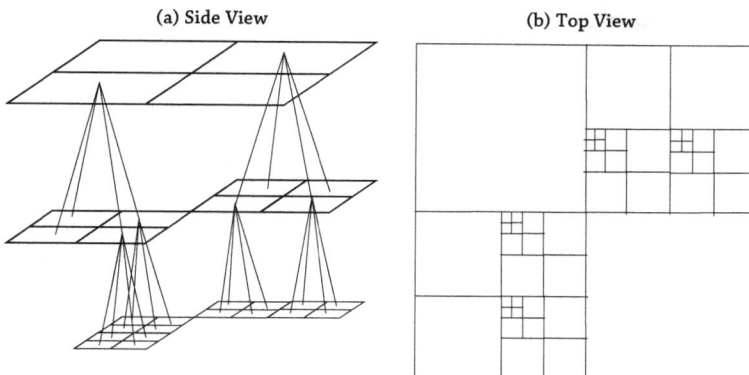

(a) Side View   (b) Top View

Figure 20.1. Quadtrees subdivide a two-dimensional space into increasingly smaller areas

Octrees implement this same idea in three-dimensional space (Kelleghan, 1997). When the world cannot be simplified into a 2D surface, the subdivision of a 3D world results in eight adjacent spaces, creating data trees with eight children rather than four (Figure 20.2).

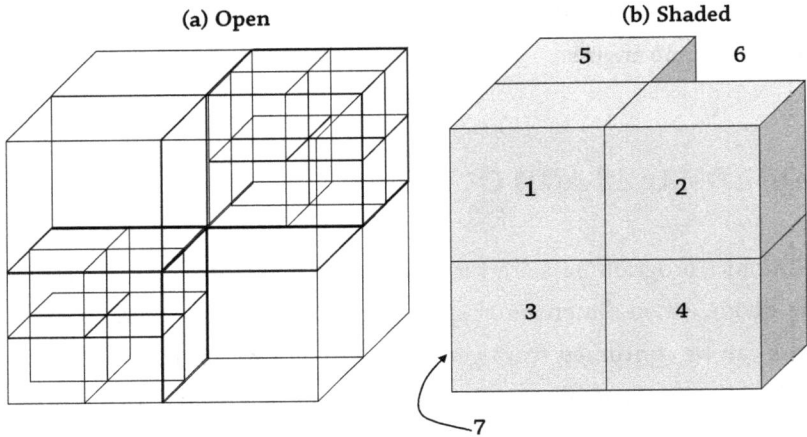

Figure 20.2. Octrees subdivide a three-dimensional area into smaller volumes

Because terrain, buildings, rivers, and forests usually reside at fixed geographic locations, managing them in a geographic grid has always been a natural approach. This has allowed the use of quad and octrees to rapidly identify the pieces of the terrain that needed to be drawn for a specific scene.

There is a close parallel to this approach for optimizing object-to-object detection for spaceships, trolls, and soldiers that are moving around in the world.

OBJECT ORGANIZATION

In this gem, object grid registration refers to the need to manage even dynamically moving objects within the same type of grid that is used for terrain data. In some cases, these objects may actually use the very same grid system.

Dynamic objects are constantly changing their locations, as well as other state variables. Individuals and groups of soldiers move from one position to another—advancing and retreating from the enemy. Therefore, from one moment to the next, the list of objects that can be seen or engaged is changing. In a brute force approach to this problem, the game engine will constantly recalculate LOS or range for every pair of objects in the game. As long as the number of objects is relatively small, this approach can be tolerated. For example, if the world contains only 10 objects fighting each other, then the number of LOS calculations is merely 90—each of the 10 objects calculates LOS to the remaining 9 objects (Figure 20.3). This is usually referred to as an order n-squared problem—$O(n^2)$. Though this solution is complete, in terms of

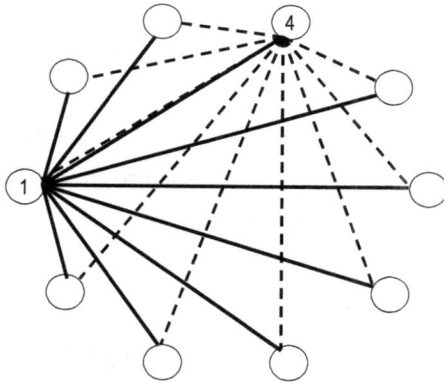

Brute Force = 10 objects * 9 detections = 90 computations
Range Pairing = 9+8+7+6+5+4+3+2+1+0 = 45 computations

Figure 20.3. Brute force object-to-object detection is an order n-squared problem

performance and scalability, it is the worst possible approach. If the number of objects increases to 100, the number of range calculations jumps to 9,900. If there are 1,000 objects in the game, it is nearly one million calculations.

One simple improvement is to implement a pairing scheme in which the range calculated from object 1 to object 4 is used immediately to determine the visibility or sensor detection from 1 to 4 and from 4 back to 1. This means that objects only need to calculate range for the other objects that are later than them in the list of objects. This cuts the number of ranges calculated in half (Figure 20.3).

These calculations may be performed each time-step if there is not some other mechanism at work. Additional filters are often used to mitigate this problem, such as keeping track of the last time that an object changed position in order to avoid unnecessarily recalculating the range between the same two positions again. But, this requires retaining some information about the previous range calculation between specific pairs of objects. This falls short of significantly reducing the complexity of the problem.

Line-of-sight is a geographic question. Arriving at an efficient solution requires a geographic approach to the problem, just as quadtrees have been used for static terrain information.

Grid Registration

Moving objects must be registered into a geographic grid as they change locations. This means calculating a grid location in addition to their more universal position. As an object moves from one grid square to another, just as a chess piece changes grid positions, it must be registered into the arriving grid and unregistered from the departing grid (Figure 20.4). The figure illustrates two lists, one for Grid(1,5) and another for Grid(1,8). At time 1234 both of these grids contain two objects. However, at time 1235, Object 1 (O1) has moved to a new position. Therefore, the grid registration process must change O1's registration to Grid(3,6). The additional computational overhead associated with managing the grid location is much smaller than the computations required to perform all of the range computations described previously. One part of this savings is due to the simpler problem of determining grid location versus the more complex calculation of range or LOS. However, a far greater portion of the savings is due to the extremely reduced number of operations that have to be performed.

Object grid registration is an order 'n' problem—O(n). LOS is an order n-squared problem—$O(n^2)$. Therefore, a game that has 100 objects in it will require at most 100 grid registrations during each time step. Without the grid (and disregarding other detection filters), these 100 objects would require nearly 10,000 range calculations each time step. So grid registration exchanges nearly 10,000 range calculations for 100 simpler registration operations.

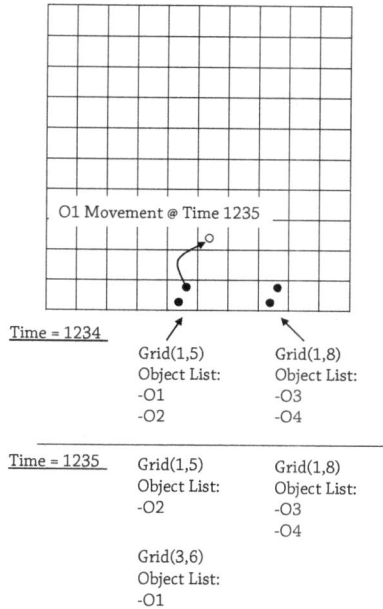

O1 Movement @ Time 1235

Time = 1234	Grid(1,5) Object List: -O1 -O2	Grid(1,8) Object List: -O3 -O4
Time = 1235	Grid(1,5) Object List: -O2	Grid(1,8) Object List: -O3 -O4
	Grid(3,6) Object List: -O1	

Figure 20.4. Grid Registration is performed each time an object's new position is in a new grid square

Weapon or Sensor Footprint Overlay

Grid registration does not entirely eliminate range or LOS calculations. It simply limits it to a much smaller number of objects. As shown in Figure 5, the sensor of the searching object overlays a small number of the grids. These grids can be identified in a number of ways, most of which are described in other articles on quadtrees and terrain database management (Frisken, 2003). Once this list of grid squares is identified, only objects that are registered within these grids need to be considered for detection or engagement. However, simply because an object is registered into one of these grids, there is no guarantee that the object also falls into the sensor footprint. In Figure 20.5 it is clear that, in many cases, the sensor does not cover all of a grid square. Objects in the partially covered grids may not actually be within range of the sensor and may need to be excluded from the list of potentially detectable or engageable objects. Therefore, a separate range or LOS calculation must still be done for the much smaller group of objects found in these grids.

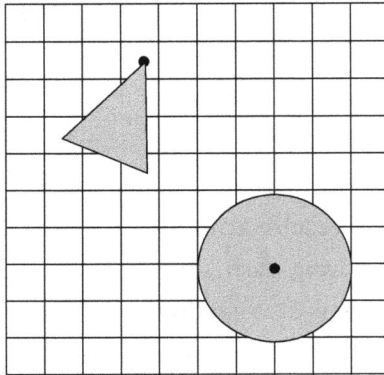

Figure 20.5. Sensor footprints overlay the grid and identify which squares are potentially in the detection field of the sensor

The programmer can implement the sensor footprint overlay algorithm such that it clearly differentiates the grid squares that are entirely within the sensor footprint from those that are only partially inside. For some games this can entirely eliminate the need for a range or LOS calculation for objects in grid squares that are entirely contained in the sensor footprint (Pritchard, 2001). However, for games where 3D terrain is an essential part of the visibility decision of AI's or NPC's, it will remain necessary to perform a separate LOS calculation for each object found in the contained grid squares. Although the grid system can identify which objects are within range of the sensor, it cannot determine whether there is a clear LOS vector from the sensor, through the terrain and obstacles, to the selected object.

Exaggerating the Footprint

Our goal is to minimize the number of objects that must be considered to be viewable by the AI agent. However, in doing this, we must also be careful that we do not inappropriately eliminate moving objects that are on the edge of the sensor footprint. Most games operate on a time step in which an object moves from point A to point B along a constant vector during a single time step. It is possible for the movement of an object to carry it across the edge of the sensor footprint such that it does not exist within the sensor footprint during either of the two discrete times at which calculations are performed (Figure 20.6a).

To capture these cases, the size of the footprint is usually exaggerated slightly. This allows the sensor to grab objects in grids that are immediately adjacent to the footprint and that may potentially move across its edge. The size of the footprint exaggeration is determined heuristically based on the size of the grid squares, the maximum speed of game objects, and the duration of the time step. The exaggeration must incorporate all grid squares from which an object can reach the edge of the footprint in a single time step. In games where movement precedes detection, the exaggeration will find the objects at their ending position and will use reverse dead-reckoning to calculate the path followed to reach that ending position and determine whether that path crossed the sensor footprint.

There are often two or three instances of footprint exaggeration. The first is an exaggeration of the size of the footprint as was just described. The second is the fact that the grid system is composed of squares and the footprint may be shaped like a wedge or a circle. The process of "squaring the circle" brings in area that is not really within the footprint. Finally, the "squared circle" will not necessarily fall exactly along grid boundaries, so it is further exaggerated to include all of the area of the "squared-in" grid squares (Figure 20.6b).

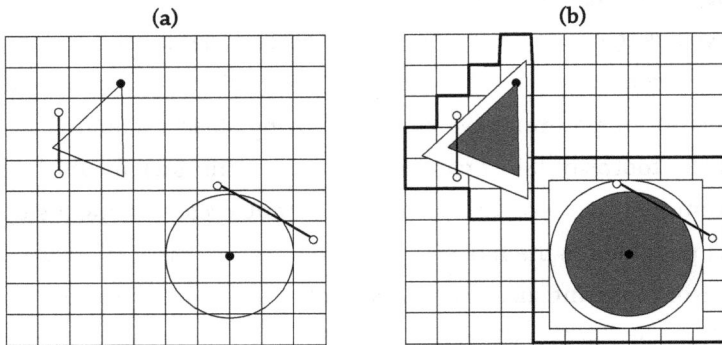

(a) (b)

Figure 20.6. Footprint exaggeration captures objects whose movement vector crosses the edge of the sensor footprint in a single time step

CONCLUSION

The method described in this gem is not complex. Once described it seems to be an obvious approach to creating a more scalable detection algorithm. However, the number of projects that continue to rediscover this approach to the problem, and other variations of it, is scandalously high. It is captured in here in an attempt to save future programmers from repeating the work of the past.

The goal is to manage dynamic object locations in a manner that is similar to that used for static terrain objects. These objects enter and leave the game, change positions frequently, and sometimes straddle two or more grid squares. All of these actions make the management of these more challenging, but the payoff in terms of saved processing time far exceeds the cost of implementing geographic grids to reduce LOS and range calculations that are one of the most repetitive calculations in a game.

Chapter 21

10 Fingers of Death

Good shooting games need good killing algorithms. This gem provides a series of combat algorithms that can be used to improve the realism of combat decisions and do so with faster algorithms. Most of the algorithms discussed here were developed for the United States military and have been validated for use in one or more real combat simulations.

First-person shooter killing algorithms are fine, but some situations can be handled more accurately and efficiently by including geometry, statistics, probability, and aggregation. Massively Multi-player and Real-time Strategy games particularly include a lot of action that does not have to be modeled with individual line-of-sight (LOS) and targeting for a headshot. These and other games can also benefit from the inclusion of multiple kill types that are based on real live-fire experiments. First-person shooters may equip AI's with some of these algorithms while leaving the more detailed LOS algorithms for the avatar controlled by the human player.

HITTING A RIBBON

The first finger of death presents a simple method for determining whether a shooter will hit a ribbon target like a road, river, long convoy of vehicles, or a serpentine creature (Figure 21.1). If the target is so long that it is impossible to overshoot or undershoot its length, the probability of hitting the target is dependent only upon the width of the target and the standard deviation of the shot pattern of the weapon. This simple algorithm is also a good way to introduce the logic and mathematics behind several of the attrition algorithms that follow (Parry, 1995). Deviations in the impact point of the munitions being fired are due to factors such as the quality of the weapon, steadiness and skill of the human operator, variations in the construction of the projectile, and wind conditions.

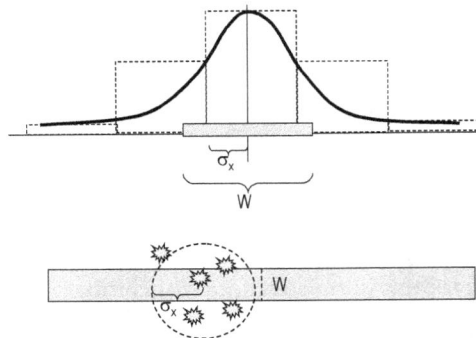

Figure 21.1. Probability of Hitting a Ribbon Target

Equation
The equation for calculating the probability of hit (P_h) for a ribbon target is:

$$P_h = 2W \big/ \sqrt{2\pi\sigma}_{\,x}$$

where,
W is the width of the target, and
σ_x is the standard deviation of the bullet dispersion in the x dimension.

This assumes that the pattern is normally distributed with the same standard deviation in both the x and y dimensions.

[Note: The code for all of these algorithms can be found on the CD-ROM.]

HITTING THE BULLSEYE

The second finger of death describes the math and probability of hitting a round target. Like the previous algorithm, this one is based on the fact that all shooters, human and machine alike, have built-in variation in every shot fired.

The algorithm is driven by two very simple variables – the radius of the target and the standard deviation of the rounds. This deviation is based on a normal distribution in which the mean value is zero because the shooter is aimed directly at the center of the target (Parry, 1995). The algorithm determines whether each shot will hit the target, but does not calculate the actual impact point of the round. This simplification eliminates calculations that would have to be done to distribute the round normally in both the x and y dimension.

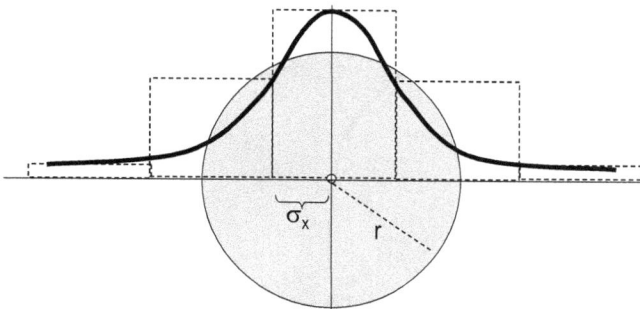

Figure 21.2. Probability of Hitting the Bullseye

Equation

The equation for calculating the probability of hit (P_h) for a round target is:

$$P_h = 1 - e^{-\left(r^2/2\sigma_x^2\right)}$$

where,

r is the radius of the target, and

σ_x is the standard deviation of the bullet dispersion in the x dimension.

HITTING A RECTANGLE

Most targets are not shaped like bullseyes, so we need a more flexible algorithm to shoot rectangular targets like the torso of a human or a vehicle. This algorithm includes measures for the length and width of a rectangular target (Parry, 1995).

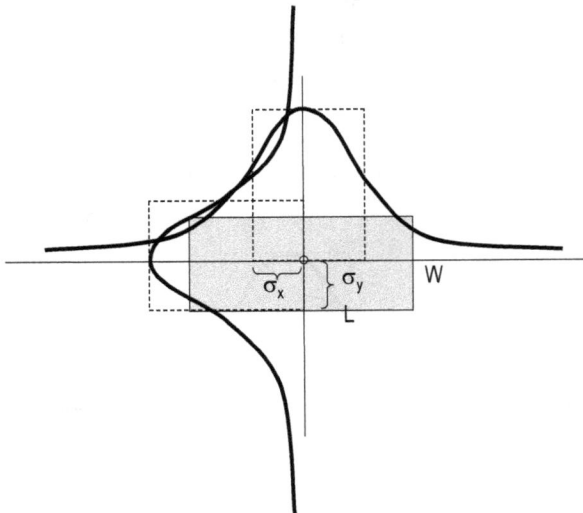

Figure 21.3. Probability of Hitting a Rectangle

Equation

The equation for calculating the probability of hit (P_h) for a rectangular target is:

$$P_h = \sqrt{A * B}$$
$$A = 1 - e^{-\left(2L^2/\pi\sigma_x^2\right)}$$
$$B = 1 - e^{-\left(2W^2/\pi\sigma_y^2\right)}$$

where,

L is the length of the target in the x dimension,

W is the width of the target in the y dimension,

σ_x is the standard deviation of the bullet dispersion in the x dimension, and

σ_y is the standard deviation of the bullet dispersion in the y dimension.

Weapons often have different standard deviations in the x and y dimensions. For example, when a football quarterback throws a pass, the variation from the aim point along the axis of flight is usually greater than the variation left or right of the aim point. The same is true for missiles being fired at a combat vehicle or rocks being thrown at a dinosaur.

SHOTGUNNING A SMALL TARGET

Some weapons unleash a barrage of rockets, bomblets, or explosive munitions all at once in an attempt to totally overwhelm the target and blow it to smithereens. When this happens, there are much faster ways of determining the killing effect of the entire barrage than calculating the impact points and lethality of each rocket individually and then accumulating them.

This algorithm calculates the probability that one of the munitions' lethal areas will overlap with the point target. The size of the target is not considered in these calculations because it is assumed that the lethal blast area can encompass an entire target (Parry, 1995).

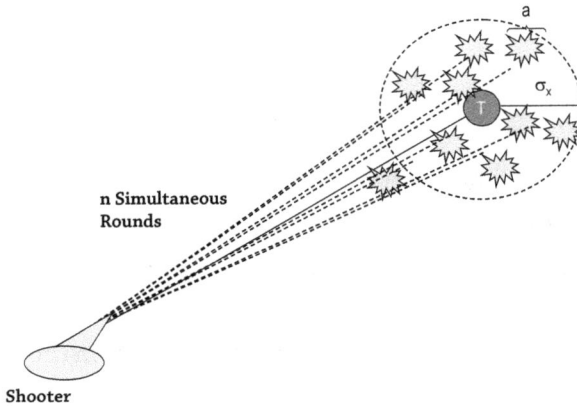

n Simultaneous
Rounds

Shooter

Figure 21.4. Probability of Killing a Target with a Simultaneous Barrage of Munitions

Equation

$$P_k = 1 - e^{-\left(na/2\pi\sigma_x^2\right)}$$

where,

n is the number of rounds in the barrage,

a is the lethal area of a single round against this target, and

σ_x is the standard deviation of the bullet dispersion in the x dimension.

DEATH BY WALKING ARTILLERY

Artillery and catapult rounds are often adjusted by a spotting team that radios corrections back to the firing battery and allows them to place the next round closer to the target. When this occurs, the lethality of the barrage is higher than the previous shotgunning method. The lethality of this walking artillery is calculated through a summation series in the exponent (Parry, 1995).

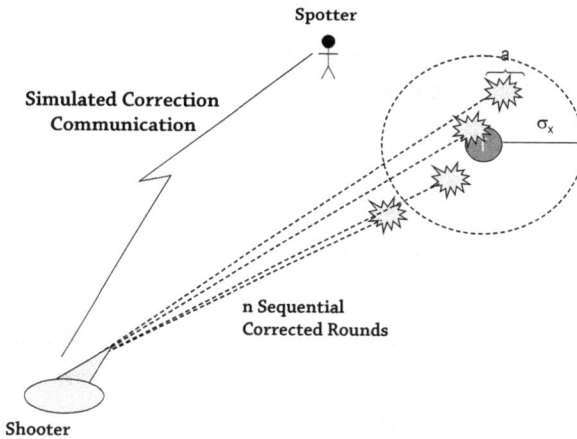

Figure 21.5. Lethality of Walking Artillery with a Spotter

Equation

$$P_k = 1 - e^{-\left(\left(a^2/2\sigma_x^2 \right) * \sum_{i=2}^{n} (i-1)/i \right)}$$

where,

n is the number of rounds fired in the barrage,

a is the lethal area of a single round against this target, and

σ_x is the standard deviation of the bullet dispersion in the x dimension.

KILLS COME IN FOUR FLAVORS

To paraphrase a famous pig, *"all kills are equal, but some are more equal than others."* Military simulations usually model four different types of kills that are most often found in real-world combat. The first flavor is a mobility kill in which the target is no longer able to move, but remains alive enough to fire its weapon or communicate with other vehicles. The second is a firepower kill in which the weapon is damaged, but the vehicle or person is still able to move. The third is a mobility *and* firepower kill in which the vehicle or person is still alive, but cannot move or use its weapon. This target may still be able to observe enemy operations, communicate, consume supplies and, in some simulations, trigger a rescue operation. The final kill type is the catastrophic kill or K-kill, often pictured as an aircraft exploding into a million pieces, a flaming tank turret spinning through the air, or a person being turned into fresh chunks of meat.

These four kill types can be pictured as a Venn diagram (Figure 21.6). Though this form clearly communicates the relationships between the kill types, in order to be applied, it has to be separated so that a specific kill type can be determined quickly for each engagement. This separated data is usually represented as a kill thermometer (Figure 21.7). Normalizing the kill types in a single space as represented in the thermometer allows a program to determine the kill type of an engagement by drawing a single random number.

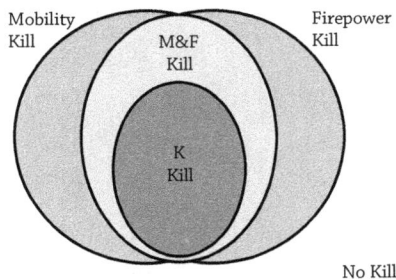

Figure 21.6. Standard Kill Types

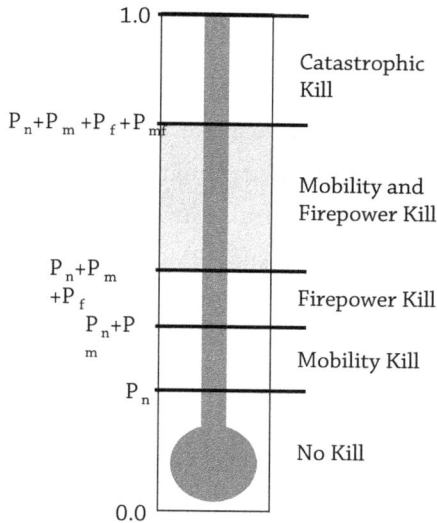

Figure 21.7. Kill Thermometer

There are live-fire projects that determine the probability of each kill type under different conditions by firing real weapons at real targets and measuring the result. Most simulations and games do not have access to such a rich source of information. Therefore, we have to identify common trends in experimental data and create equations that mimic those while remaining flexible enough to be applied to new weapon/target pairs. One simulation project noticed a distinct relationship between the mobility, firepower, and catastrophic kill data they had received from live-fire experiments. This relationship allowed them to create simple equations that use the root of a single "probability of mobility or firepower kill" (P_{MoF} is the union of all of the shaded areas above) to calculate all of the other probabilities. The trend they noticed was that a mobility kill occurred 90% of the time that damage was done ($P_M = 0.9^* P_{MoF}$); a firepower kill occurred 90% of the time ($P_F = 0.9^* P_{MoF}$); and a catastrophic kill occurred 50% of the time that damage was done ($P_K = 0.5^* P_{MoF}$).

However, this information cannot be applied directly to the kill thermometer in Figure 21.7. P_M does not say that 90% of all engagements result in a mobility kill. It says that 90% of mobility-or-firepower kills include a mobility kill. Therefore, it has to be separated to make it possible to draw a single random number and determine which kill to apply to the target. These independent kill probabilities can be extracted as shown below.

Equation

$$P_n = 1.0 - P_{MoF}$$
$$P_m = P_{MoF} - P_F = 0.1 * P_{MoF}$$
$$P_f = P_{MoF} - P_M = 0.1 * P_{MoF}$$
$$P_k = 0.5 * P_{MoF}$$
$$P_{mf} = P_{MoF} - P_m - P_f - P_k = 0.3 * P_{MoF}$$

where, the small subscript indicates the probability that only one type of kill occurs. For example, P_m is the probability of *only* getting a mobility kill, but not getting any other form of kill. P_n is the probability of no kill occurring.

These independent kill probabilities determine where the breakpoints fall in a kill thermometer and can be easily programmed as shown in the code on the CD-ROM.

CHEMICALS, FIREBALLS, AND AREA MAGIC

There have been many models of the dispersion of chemicals and other agents. The following simple algorithm calculates the probability of a kill based on the volume of chemical released and the distance that the release occurs from the target. For games, this algorithm could be used for expanding fireballs, area magic, or any other exotic and evil weapon.

Equation

$$P_k = \left(\sqrt[3]{nw_r} \big/ \sqrt{2\pi} \right) * e^{-0.5*\left(k*d^2/nw_r\right)^2}$$

where,

n is the number of rounds impacting at a specific point,

w_r is the weight of the chemical inside of each round (in kilograms),

d is the distance that the rounds fall from the target location (in meters), and

k is a constant representing the dispersion characteristic of the chemical. For these experiments we recommend beginning with a value of 0.00135.

This equation allows you to deal with each round individually or to aggregate multiple rounds into a single attack centered at the same impact point. The equation also incorporates the constant k that represents the density and viscosity of a chemical compound. You can adjust this value to create the effect desired.

THE SHRAPNEL WEDGE

When an aircraft is shot down with a missile it is seldom accomplished by the missile flying directly into the aircraft. More often, the missile reaches a "point of closest approach" and explodes near the aircraft. The shrapnel from the missile then spreads out in a donut or spherical pattern from the point of explosion and hopefully, the aircraft is caught in that shrapnel pattern and destroyed (Ball, 1985). This algorithm can be used with exploding projectiles, fireballs, and magic targeted at aircraft, dragons, and spacebugs.

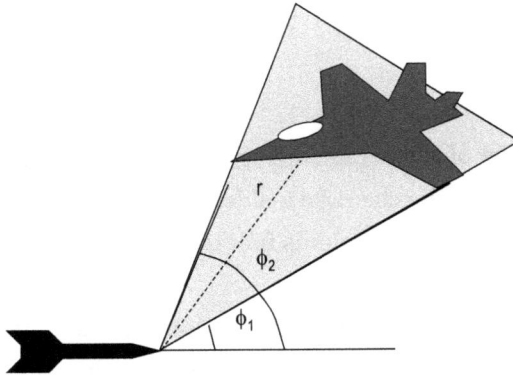

Figure 21.8. Probability of Killing a Target in a Shrapnel Wedge

Equation

$$x = nA_v / \left(2\pi r^2 \left(\cos\phi_1 - \cos\phi_2\right)\right)$$
$$P_k = 1 - e^{-x}$$

where,
n is the number of fragments or projectiles in the missile warhead,
A_v is the vulnerable area of the target presented to the missile (in sq meters),
r is the range from the detonation point to the target (in meters),

ϕ_1 is the angle from the trajectory of the missile to the near edge of the vulnerability area of the target, and

ϕ_2 is the angle from the trajectory of the missile to the far edge of the vulnerability area of the target.

BEATING THE BUSHES

Some engagements involve teams of hunters searching the terrain or bushes for hidden prey (Shubik, 1983). When a large group of hunters is looking for a large group of prey, it is possible to model the capture or kill of the prey in an aggregate form, rather than representing the individual movement and line-of-sight of every hunter and every prey. As before, this approach is very valuable when the hunting and killing is being conducted by AI controlled hunters and especially when it is happening off the player's screen.

The algorithm is structured to calculate the change in the population of the prey based on the number and efficiency of the hunters. It also accounts for different types of prey and hunter animals, e.g. small rodents, medium-sized wolves, and large elephants.

To use the algorithm, we must define a probability of detection for each type of hunter against each type of prey under the given conditions (open terrain, forest, city, etc.). We must also select a "hardness" factor that differentiates the ability of the prey to elude, escape, or survive the actions of the hunter. These numbers are usually determined heuristically through experimentation and observation.

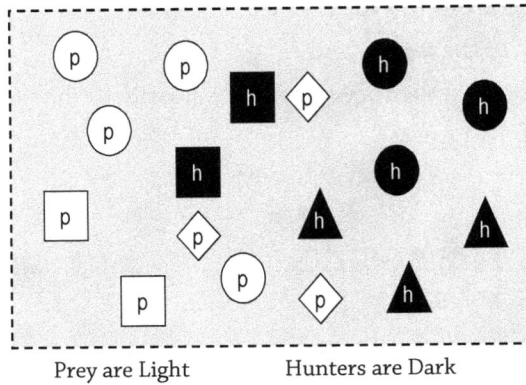

Prey are Light   Hunters are Dark

Figure 21.9. Multiple Types of Hunters Searching for Multiple Types of Prey

Equation

$$x = \left((k_j / p) * \sum_{i=1}^{n} D_{i,j} * h_i \right)$$

$$A_j = p_j * (1 - e^{-x})$$

where,

A_j is the number of kills of animal type j,

p_j is the number of prey of type j,

k_j is a hardness measure of the prey in the range [0,1],

p is the total number of prey of all types,

n is the number of prey types

$D_{i,j}$ is the probability that a hunter of type i can detect a prey of type j, and

h_i is the number of hunters of type i.

BEATING THE BUSHES WITH PREY SPACING

The final finger of death is a modification of the previous one. Mathematicians and analysts noticed that the previous algorithm did not account for differences in the density of prey hiding in the bushes. It is clearly much easier to find and kill prey when there are a hundred of them in the search area than if there are just two or three. Therefore, they created a variation known as the Lulejian model (Shubik, 1983) in which the spacing between the prey is an important factor. The visual picture for this algorithm is the same as that above, but the mathematics differ to account for the spacing of prey. The definition of k_j also varies slightly in that Lilejian defines k_j as the average destruction of the hunters on prey type j.

Equation

$$x = \left(\sum_{i=1}^{n} k_j * h_i \right) \Big/ (s * p)$$

$$A_j = p_j * (1 - e^{-x})$$

where,

A_j is the number of kills of prey type j,

p_j is the number of prey of type j,

s is the average spacing between the prey in the search area (in meters),

p is the total number of prey of all types,

n is the number of prey types

k_j is the average destruction of the hunters on prey type j, in the range [0,1],

h_i is the number of hunters of type i.

CONCLUSION

The ten fingers of death described in this chapter are just a few of the combat killing algorithms that can be applied to computer games. The concepts of geometry, probability, statistics, and physics used in the ten fingers of death are good examples of approaches to many problems. Game developers should do what military modelers do to improve these—apply experience, mathematics, creativity, and other sciences to find equations that work well for your game. Don't be afraid to experiment!

Chapter 22

Synchronizing a Multiplayer Server Farm

*S*ynchronization of event execution and time progression across parallel and distributed computers has been an issue in high performance computing circles for decades. The programmers who created massive models of nuclear blast effects found that their simulations were too big to be handled by a single processor of any size. Therefore, they turned to parallel computing to give them the horsepower to run these in a reasonable amount of time. But that immediately created a new problem of synchronizing the events that were occurring on these multiple independent processors.

Since then, distributed computing has spread to a number of other application domains. Interactive training for the military and computer gaming are two of the most popular. Massively multiplayer games (MMPGs) are the most extreme form of distributed computer game and demand event synchronization just as the nuclear models and training simulations do.

IMPACTS ON THE MASSIVELY MULTIPLAYER EXPERIENCE

MMPGs like Asheron's Call find that event and time synchronization across machines in their server farm is a major issue. Jeff Johnson of Turbine Entertainment says that, "the biggest problem that Turbine has in managing its seamless worlds' server-side is in dealing with asynchronicity and serializing server game state at arbitrary points of execution" (Johnson, 2004). Generally MMPG server systems follow one of two major architectures: (1) the zone-based model where objects are represented on a single machine that is responsible for a specific geographic area of the virtual world, and (2) the distributed (or seamless) model in which objects are represented on multiple servers and their state values and event lists must be constantly synchronized (Beardsley, 2003). Though the second model desperately requires a reliable synchronization mechanism, even the zone-based model contains game events, boundary conditions, and system management operations that require synchronization.

Poor or non-existent synchronization across servers impacts both the players' experience with the game and the ability to manage and control operations within the game engine. In some cases, inconsistent event execution can result in different outcomes of event sequences on the servers and potentially expose this inconsistency to the player client machines.

Synchronization is a difficult thing to achieve and does impose performance costs and operational limitations. However, the increasing complexity of MMPGs and the growing horsepower and bandwidth that drives them is going to justify the resources for synchronization just as these resources have opened the door for better AI and physics in the past.

AVAILABLE SOLUTIONS

There are a number of different solutions to the event and time synchronization problem. Each provides slightly different capabilities at different performance costs. The first, most common, and least expensive is the *best effort* method. This calls for each message receiving process to buffer messages, order them according to their timestamp, and execute them in the hope that all of the sent events for the buffered period have been received. When a late message is received, the event managing software must make a decision to either execute the event late or to delete it. This decision is usually very specific to the game and the type of the event. It is essential that this decision is made deterministically and applied identically on every machine. However, even a deterministic algorithm cannot deliver uniform results on multiple machines. Message delivery delay varies from one machine to the next and the buffering of events does not result in the same events being included in the buffers on every machine. For example, consider an event E1 sent to computers M1 and M2. On computer M1, event E1 may have been received, ordered, and properly executed. But on computer M2, event E1 may have arrived late and been subject to either late execution or deletion. This uncertainty of results is a major motivation for the creation of more predictable synchronization algorithms.

A second popular method of synchronization is through the use of a *central timeserver*. One process is anointed as the master of all time progression. Its job is to set the pace of execution for all game processes in the server farm and to determine when conditions warrant moving forward, slowing down, or stopping. This method improves on *best effort* in that all of the processes are slaved to one master and thus remain much more closely aligned in time. However, it does not provide any mechanism to guarantee that messages are executed in the same order on multiple machines. Also, as the "master of time", this process can ignore the performance issues of heavily loaded slave processes, allow them to fall behind, and create opportunities for causal event violations. Many of the message passing algorithms that have been created to address this problem are actually ad hoc or partial implementations of synchronization algorithms we are about to prescribe.

The leading method for reliable event and time synchronization is the *Chandy/Misra/Bryant* (CMB) algorithm and several useful modifications of this algorithm. CMB requires exchanging messages between servers that define which event timestamps have been executed and determine the readiness to move forward to the next time increment (Fujimoto, 1990). In this chapter, we will describe CMB and some modifications that are particularly useful to MMPGs.

Another synchronization method that is very popular within academia and some analytical communities is Time Warp. This is a very exotic method that is difficult to understand, implement, and modify for an MMPG. Readers interested in this technique should dig into the references provided at the end of the chapter (Fujimoto, 2000).

TIME IN THE VIRTUAL WORLD

Before explaining CMB event and time synchronization in greater detail, it is important to establish some basic properties of time management in simulations and games. Some of these characteristics are required to create a simulation that is causally consistent and others are necessary to enable CMB to work.

1. ***Virtual Time is Real.*** When discussing the subject of time in a virtual world, the significant value is the time that is created and managed by

the software, not the "real time" experienced by flesh-and-blood play-ers. When we manage time advance in a game we are often attempting to align virtual time with real time, but all events and the entire digital world are referenced to virtual time, not real time.

2. **Discrete Step Size.** In a game, time moves forward in discrete incre-ments. In many systems these steps are so small that it appears to the player that time is moving continuously. But that is an illusion just as a movie appears to present a continuous moving image even though it actually has a step size of 24 frames-per-second. Effectively, the step size defines an increment that is small enough that all events sched-uled in that period can be treated as if they occurred simultaneously.

3. **Monotonically Increasing.** Game time is always monotonically in-creasing. This means that event timestamps always increase or stay the same. They do not ever go backward. This is an important property because it means that if a process generates an event with a timestamp of 100 on it, then it will never again generate an event with a stamp of 99, 98, or any other value less than 100.

4. **Event Timestamps.** All events in a simulation are time stamped. There are no orders, commands, inquiries, or reports that are created without a stamp indicating the time at which they should be executed. It is not so important to stamp them with the time that they are cre-ated, though there is some value in that as well, the essential time is that at which the event must be executed by the game.

5. **Network Message Lag.** The delivery of messages across a computer network always takes time and induces lag. Therefore, messages on the receiving end are "old" in that they represent the state of the send-

er some delta milliseconds in the past. Additionally, there is no guarantee that messages sent will be received in the order that they were sent, or even received at all.

6. **Limited Remote Information.** The receiver of messages always has a limited amount of information about the state of the sender of the messages. This limitation has a direct impact on the content of messages and the reasoning that must be performed on the receiving computer.

MANAGED SYNCHRONIZATION

It is possible to provide much more reliable synchronization than the Best effort described above. Below are several of the leading solutions for event and time synchronization.

Chandy/Misra/Bryant Algorithm

CMB is known as a "distributed k-reduction algorithm". It can be used to synchronize any number of independent processes running on different processors or on the same processor. Using event timestamps from all of the participating processes the CMB algorithm calculates the "Global Virtual Time" (GVT) for the entire group. GVT is the minimum timestamp on all exchanged messages. Virtual worlds generate and transmit events that are scheduled to be executed at some time in the future. GVT identifies the latest timestamp on events that can safely be processed without creating a causal error. All events up to and including those at GVT can be processed without worry that a synchronized process will create another event in the past of GVT.

When implementing CMB, the infrastructure that receives the event messages maintains one queue for each of the other remote participants in the synchronization. As events arrive from the remote processes, they are logged in the appropriate queue (Figure 1). The GVT mechanism evaluates the time-stamps in each queue and identifies the lowest value in the queues. That time-stamp becomes the next GVT value and all events with that timestamp are released to the modeling software for execution. Events with higher time-stamps remain in the queue awaiting future release. As events are released, it is possible for one or more of the queues to become empty. When this occurs the mechanism cannot advance the value of GVT because it cannot determine what the lowest timestamp for the associated process will be. Therefore, when this occurs, GVT must remain at its current value until an event arrives to fill the empty queue. This algorithm is illustrated in the following code sample (Perumalla, 2004).

```
/* basicGVT illustrates the original Chandy/Misra/Bryant algorithm – with-
out mods */
void basicGVT( void )
{
/* integer used for iterating through the list of processes */
int processIndex = 0;

/* integer used for iterating through the event queue of each process*/
int queueIndex = 0;

/* Initialize stop */
        bool stop = NULL;

/* Next Global Virtual Time "nextGVT" initialized to some large number */
nextGVT = setNextGVT (99999);

/* This while loop iterates through each of the processes and their queues,
and checks for the lowest timestamp.*/
while ( ((processIndex < numberOfProcesses) && (stop == NULL)) ; pro-
cessIndex++)
{
        /* If event queue is not ordered, check every
event. If ordered, just check the first event. This for loop iterates until the
queue is empty */
```

```
for (queueIndex = 0; e[processIndex][queueIndex]; queueIndex++)
{
/* Checks the timestamp to see if it is the new minimum and sets nextGVT
if it is. */
nextGVT = min(nextGVT, e[processIndex][queueIndex]->timestamp);
        } /* end for */

/* If the current "processIndex" queue was empty. */
        if (queueIndex ==0)
{
/* stop calc*/
                stop = 1;

/* Sends a NULL message.  Null messages are required to keep the process
from entering a deadlock state. */
                sendNullMessage ( );

        } /* end if */

} /* end while */

if (stop == NULL)
{
        GVT = nextGVT;
} /* end if stop */

} /* end basicGVT */
```

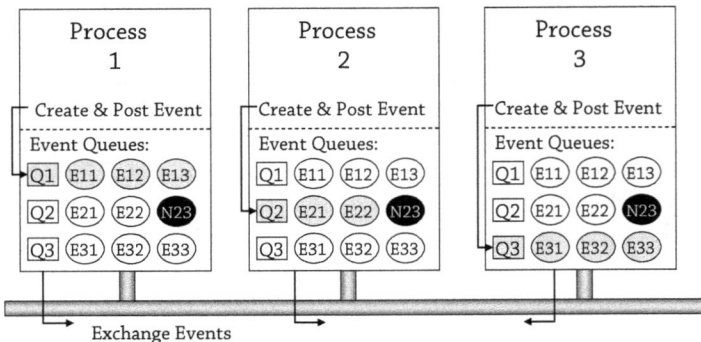

Figure 22.1. Chandy/Misra/Bryant Event Queues to Calculate GVT

Empty queues can result in a deadlock in which process P1 is awaiting an event from process P2, while P2 is waiting for an event from P3, and P3 is awaiting an event from P1. To break this deadlock, CMB implements "Null Messages" (the black messages in Figure 22.1). These are not true executable events. They are messages that simply carry the timestamp for the next event that a process intends to generate. Null Messages are usually generated at a scheduled rate that is driven by the largest acceptable deadlock time. The original CMB algorithm generated a Null Message after each real executable event. This essentially reduced deadlock time to zero, but at the cost of doubling the number of messages being sent between computers. The newer timed-release mechanism is much more bandwidth economical, but at the expense of a short deadlock period.

The computation of GVT is so simple that the CPU expense is almost insignificant. The real impact is that it regulates the pace of all processes to match that of the slowest process in the synchronization. This is one of the features that later modifications have improved upon.

Advance Request/Grant Modification

CMB was designed for analytical simulations such as nuclear blast studies or models of national air traffic patterns. When this algorithm migrated into the military training domain, specific modifications were made to improve its performance. In a military training simulation, there can be thousands of objects sending event messages to dozens of different computers. Evaluating the timestamp on all of these messages proved to be redundant and unnecessary. During a given time step, thousands of objects generate event messages with the same timestamp. Under these conditions, the GVT algorithm found itself comparing thousands messages with the same timestamp.

To reduce these comparisons, an "Advance Request" and "Advance Grant" message pair was created. Advance Request was a request by the process to move to a specific time in the future (Figure 2). In most cases, this corresponded to the timestamps on the thousands of event messages. But, it reduced the number of timestamps to be compared for GVT from a number on the order of the number of objects in the virtual world (n*10,000s), to a number on the order of the number of processes running (n*10s). In the case of large object database this can represent an improvement of three orders of magnitude. When the GVT algorithm determines the next safe timestamp to advance to, it provides an "Advance Grant" message to those processes that are allowed to move to their requested time. Processes that had requested a time further in the future do not receive a response and are expected to wait until they receive a grant—usually after a slower machine has caught up to that time.

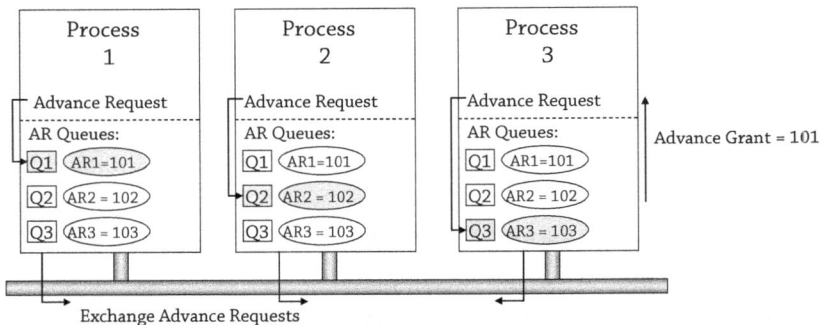

(Distributed events are not shown in these queues. They are handled separately.)

Figure 22.2. Advance Request Messages Reduce Cost of GVT Calculation

The changes to the original method shown above are limited to the while loop. In this snippet of sample code it is clear that the number of events being evaluated has been significantly decreased through the elimination of the entire inner for loop.

```
/* This while loop iterates through all of the processes, and looks for the
advanceRequest message with the lowest time. It then sets the nextGVT to
that time. Note that this algorithm only checks each process once for an
advance request rather than iterate through each processes entire queue. */
while ( ((processIndex < numberOfProcesses) && (stop == NULL)) ; pro-
cessIndex ++)
{
/* Checks the process for an advance request */
        if (e[processIndex]->advanceRequest)
{
/* sets the nextGVT to the advance request time if it is less than the current
minimum*/
                nextGVT = min(nextGVT,
e[processIndex]->advanceRequest);
}
/* Sends a null message if the queue is empty */
else
{
/* stop calc*/
stop = 1;

/* Sends a NULL message. Null messages are required to keep the process
from entering a deadlock state. */
sendNullMessage ( );

        } /* end if */

} /* end while */

if (stop == NULL)
{
/* Sets the Global Virtual Time "GVT" to the new time obtained from the
advance request */
        GVT = nextGVT;

/* Publish Advance Grant message. Once this is published processes are al-
lowed to move to the requested time and execute events for those times.*/
sendAdvanceGrant(GVT);

} /* end if */
```

This modification significantly improved the performance of calculating GVT. However, it did not improve the situation in which the slowest process was regulating the entire family of processes involved in CMB.

Lower Bound Time Stamp and Lookahead

The next major modification to CMB is often referred to as the Lower Bound Time Stamp (LBTS) method [Mattern93]. This takes advantage of the fact that most simulations and games have a defined, discrete time step size. Under the traditional GVT method, when one process is operating on events at time 100, other remote processes are allowed to process all events up to and including those with stamp 100. However, using LBTS, remote processes recognize that a simulation operating at 100 right now will generate future event messages with timestamps of 100 plus one time step. Therefore, if a process' step size is 4, a remote simulation can be given permission to execute all events up to and including those with stamps of 104, knowing that the first process will not generate a message at 101 because it is not capable of it. Under the exact same conditions, LBTS is more aggressive than GVT and allows faster processes to move ahead of slower ones by some fraction of one step size. This can be extremely useful when different processes use different step sizes. In some cases, there are simulations with a step size of 1 working together with others using a step size of 2 or 3. (These are conceptual numbers that illustrate the ratio of size. An actual simulation process would use a step size such as 100 milliseconds, 200 milliseconds, or 1 second—which have ratios 1:2:10 and can be exploited by LBTS.) When this happens, the simulation using step size of 1 will find useful work to do with stamps or 103, when the simulation with step size of 2 would otherwise have regulated it back to 102 under GVT (Figure 22.3).

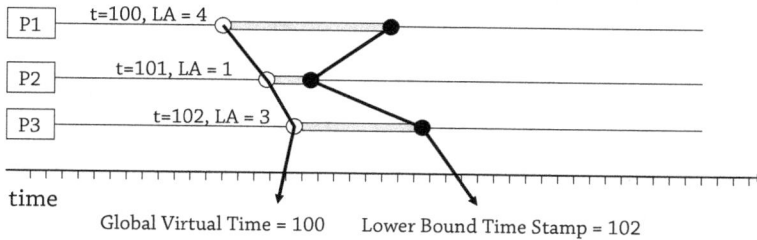

Figure 22.3. Comparison of GTV and LBTS

This slight modification can maintain causal consistency across the family of servers while also reducing the drag caused by the slower machines. The changes necessary to implement this are entirely limited to the `if` statement on `advanceRequest`. GVT is no longer equal to the lowest advance request value, but to the lowest of the sum of each processses' advance request and its lookahead value.

```
if (e[processIndex]->advanceRequest)
{
    nextGVT = min(nextGVT,
    (e[processIndex]->advanceRequest + e[processIndex]->lookAhead));
}
```

Deconflicting Simultaneous Timestamps

The discussion above provides the general solution to the synchronization problem. But there are several unique issues that must be dealt with to allow this mechanism to work. In a multiserver game, it is desirable to have all events processed in the same order on each server. This goes beyond time ordering them. It includes processing those with the same timestamps in the same order on multiple machines. Several interesting mechanisms have been created to support this (Fujimoto, 2000).

Assume that the game time step size is 100 milliseconds. Then a time-stamp value would have one hundred milliseconds as their smallest digit. A timestamp value of "12345" would represent 1,234 seconds and 500 milliseconds into the game. To support absolute ordering of events, this number must be augmented with more information. Some researchers point out that this is almost equivalent to stamping events with a unit smaller than the game's native step size. But, there other techniques that are a little more complex than that.

The first option is to add an ID number to the timestamp that is an incrementing counter that resets to zero each time the game ticks to the next step. This means that every event is tagged with a time, such as 12345, but also receives a counter. Therefore its time and ID number might be 12345.002, followed by an event stamped 12345.003, 12345.004, etc. This inclusion of an ID number works very well for indicating the order in which events are created. It also allows a receiving process to use this information to identify any event messages that are missing. In this paper we use '.' to delimit the information, which is a useful method for explaining the concepts. But in practice these values may fit into different digit positions within a single large integer or may be stored in different variables.

However, the order in which messages are created is not necessarily the order in which they should be executed. This has led to the practice of creating an "age" and a "priority" for messages. The "age" identifier is like a generational indicator. Events that are stored in the initial starting data set have an age of 0. When one of those events causes another event to be created, the created event has an age of 1. When that one causes an event, it will have an age of 2. If an event of age 3 triggers the creation of two new events, then both of them have an age of 4. This insures that whenever an event is caused by another event, the causal order between the two is maintained. When age is combined with the unique ID described earlier, it insures that an ordering algorithm always places two sibling events in the order that they were created. Adding age as part of the timestamp can be done in many ways, but we will illustrate it as "timestamp.age.ID" or "12345.002.004" in which 12345 is the actual timestamp, 002 is the age, and 004 is the unique ID.

Priority indicates specific events that should be processed ahead of others. These are usually events that have a need to be executed very quickly after being sent. For example, in an FPS game, any explosion events should have a higher priority than player-to-player chat messages. The use of a priority stamp can replace the use of age or be combined with it. Retaining all of these pieces of data may result in a timestamp that includes "timestamp.priority.age.ID".

The age and ID modifications were created to improve synchronization within a time step. Priority allows the sender to specify which events should be addressed first.

CONCLUSION

This chapter has described algorithms that are designed to synchronize event execution and time advance in distributed simulations and virtual worlds. These techniques have been in use for many years in the high performance computing community. As MMPGs grow more complex and comprehensive, they develop a similar need for strict synchronization between some or all of the servers within the server farm. The Chandy/Misra/Bryant algorithm and the modifications shown here are the most applicable event and time synchronization algorithms that can be applied in this environment. At one time the speed of computers and networks limited the use of these algorithms to simulations with timestep sizes on the order of 1 minute. However, as hardware performance has improved, it has been possible to bring these algorithms into simulations operating with 1 second or 100 millisecond timesteps. Continually improving hardware performance, algorithm optimization, and the complexity of distributed virtual worlds will make these methods accessible to simulations with even smaller timesteps in the future.

Like all new additions to game software, the computational costs and impacts of the new algorithm must be balanced against the benefits provided. Many games and even military simulations operate sufficiently well without strict synchronization. The purpose of MMPGs is to provide a believable immersive experience. The demands for accuracy are not as high as those for modeling nuclear blasts or chemical reactions. Algorithms like CMB will earn their way into MMPGs as they become better understood, the costs for implementing them are known, specific MMPG optimizations emerge, and the complexity of MMPG worlds increases.

REFERENCES

Abbott, Edwin. 1884. *Flatland: A Romance of Many Dimensions*. 1983 Reprint. Barnes & Noble. New York, NY.

Abt, C. (1970). *Serious Games*. New York: The Viking Press.

Allen, T.B. (1987). *War Games: The secret world of the creators, players, and policy makers rehearsing World War III today*. New York: Berkeley Books.

Amazon.com. (2006). "Amazon Elastic Compute Cloud (EC2)". Amazon.com web site accessed at http://www.amazon.com/gp/browse.html?node=201 590011

Anderson, C. (2006). *The Long tail: Why the future of business is selling less of more*. New York: Hyperion Books.

Anderson, C. (October 2004). "The Long tail". *Wired Magazine*, 12(10). http://www.wired.com/wired/archive/12.10/tail.html

Aoyama, Y. & Izushi, H. (2003). "Hardware gimmick or cultural innovation? Technological, cultural, and social foundations of the Japanese video game industry". *Research Policy*, 32.

Armour, F. and Kaisler, S. (2001). "Enterprise architecture: Agile transition and implementation". *IT Professional*, 2(3), pp. 30-37.

Armour, F.; Kaisler, S.; and Liu, S. (1999). "A Big-picture look at enterprise architectures". *IT Professional*, 1(1), pp. 35-42.

Ashley, S. (February 20, 2006). "Cognitive radios". Scientific American online.

Ball, Robert E. (1985). *The Fundamentals of Aircraft Combat Survivability Analysis and Design*. AIAA Press.

Banks, J. Editor. 1998. *Handbook of Simulation: Principles, Methodology, Advances, Applications, and Practice*. New York: Wiley-Interscience.

Barnes, N. (January 2005). "The Restructuring of the retail business in the US: Fall of the shopping mall". *Business Forum*, 27(1).

Bauer, J.M. (June 1997). "Market power, innovation, and efficiency in telecommunications: Schumpeter reconsidered". *Journal of Economic Issues*, 31(2), pp. 557-565.

BBC. (Nov 18, 2004). "Remote control rifle range debuts". British Broadcasting Company News. Accessed at http://news.bbc.co.uk/2/hi/technology/4022147.stm

Beardsley, Jason. (2003). "Seamless Servers: The Case For and Against", *Massively Multiplayer Game Development*, Charles River Media.

Beck, J. and Wade, M. (2004). *Got Game: How the gamer generation is reshaping business forever*. Boston, MA: Harvard Business School Press.

Ben-Natan, Ron. 1995. *CORBA: A Guide to Common Object Request Broker Architecture*. McGraw Hill. New York, NY.

Bergeron, B. (2006). *Developing serious games*. Charles River Media.

Bloch, P.; Ridgway, N.; and Dawson, S. (January 1994). "The Shopping mall as consumer habitat". *Journal of Retailing*, 70(1).

Boar, B. (1999). "A Blueprint for solving problems in your IT architecture". *IT Professional*, 1(6), pp. 23-29.

Bodansky, Yossef. (2001). *Bin Laden: The Man Who Declared War on America*. Prima Publishing. Roseville: CA.

Boorstein, J. and Watson, N. (September 15, 2003). "40 Under 40: Young, Powerful, and Changing the World". *Fortune Magazine*. http://www.fortune.com/fortune/forty/articles/0,15114,479712,00.html

Booz-Allen & Hamilton. (1997). "Critical Infrastructure Protection Strategic Simulation Report". Report to the President's Commission on Critical Infrastructure Protection. Washington D.C.

Boswell, B. (Spring 1998). "Time to market". *Evolving Enterprise* at http://www.lionhrtpub.com/ee/ee-spring98/boswell.html

Brockman, J. (not dated). "Nathan Myhrvold: The Chef". Digerati web site. http://www.edge.org/digerati/myhrvold/myhrvold_p1.html

Brunn, C. (December 2003). "The Economy as an agent-based whole—Simulating Schumpeterian dynamics". *Industry and Innovation*, 10(4), pp.475-491.

Bushnell, N. (August 1996). "Relationships between fun and the computer business". *Communications of the ACM*, 39(8).

Caniff, M. (1992). *War games: Two complete Steve Canyon adventures*. Amherst, MA: Kitchen Sink Press.

Carella, C. (July 7, 2006). "Home run derby broadcast live in SL". *The Daily Graze*, online at http://blogs.electricsheepcompany.com/chris/?p=92

Carr, N. (July/August 2008). Is Google making us stupid? *The Atlantic.com*. http://www.theatlantic.com/doc/200807/google

Carr, N.G. (2004). *Does IT Matter? Information Technology and the Corrosion of Competitive Advantage*. Boston: Harvard Business School Press.

Carr, N.G. (May 2003). IT Doesn't Matter. *Harvard Business Review*. Boston: Harvard Business School Press.

Carr, N.G. (May 27, 2004). The Z Curve and IT Investment. *Digital Renderings*. Nicholas G. Carr web site, accessed at http://www.nicholasgcarr.com/digital_renderings/archives/the_z_curve_and_it.shtml

Carroll, J. and Cameron, D. (June 2005). "Machinima: Digital performance and emergent authorship". *International DiGRA Conference*. Accessed at http://www.gamesconference.org/digra2005/papers/a3f603e92ab2d55201bc2e c45509543.doc

Casti, J. (1997). *Would-be Worlds: How simulation is changing the frontiers of science*. New York: John Wiley & Sons.

Chandler, A. (2001). *Inventing the electronic century: The Epic story of the consumer electronics and computer industries*. New York: Free Press. *Change Management*

Chatham, R.E. (July, 2007). Games for training. *Communications of the ACM*, 50(7), 37-43.

Chesbrough, H. (2003). *Open innovation: The New imperative for creating and profiting from technology*. Boston, MA: Harvard Business School Press.

Christensen, C. (1992). *The innovator's challenge: Understanding the influence of market environment on processes of technology development in the rigid disk drive industry*. Doctoral Dissertation. Harvard Business School, Boston, MA.

Christensen, C. (1997). *The Innovator's dilemma: When new technologies cause great firms to fail*. Boston, MA: Harvard Business School Press.

Christensen, C. (2004). "Exploring the limits of the technology S-curve. Part II: Architectural technologies" in *Strategic management of technology and innovation* by R. Burgelman, C. Christensen, and S. Wheelwright, Boston, MA: McGraw Hill.

Christensen, C. and Raynor, M. (2003). *The Innovators solution: Creating and sustaining successful growth.* Boston, MA: Harvard Business School Press.

Coleman, M. (2003). *Playback: From the Victrola to MP3, 100 years of music, machines, and money.* Cambridge, MA: Da Capo Press.

Conger, J.; Spreitzer, G.; & Lawler, E. (1999). *The Leader's change handbook: An Essential guide to setting direction and taking action.* San Francisco: Jossey-Bass Publishers.

Coram, R. (2004). *Boyd: The fighter pilot who changed the art of war.* New York: Little, Brown, & Co.

Csaba, F. and Askegaard, S. (1999). "Malls and the orchestration of the shopping experience in a historical perspective". *Advances in Consumer Research,* 26.

Cutler, G. (April 22, 2007). Coalition Casualty Count Metrics. http://www.icasualties.org/

Daily, L. (October 2006). "But mom, this is my homework". *Delta Sky Magazine.* http://www.delta-sky.com/2006_10/RolePlaying/index.html

Davila, T.; Epstein, M.J.; & Shelton, R. (2006). *Making innovation work: How to manage it, measure it, and profit from it.* Philadelphia, PA: Wharton School Publishing.

Davis, P.K. (August 1995). "Distributed Interactive Simulation in the Evolution of DoD Warfare Modeling and Simulation". *Proceedings of the IEEE.* 83(8), pp. 1138-1155.

Defense Modeling and Simulation Office. 1995. Department of Defense High Level Architecture for Modeling and Simulation, Version 0.2. DMSO. Alexandria, VA.

Defense Modeling and Simulation Office. 1996. High Level Architecture—Object Model Templates. DMSO. Alexandria, VA.

Defense Modeling and Simulation Office. 1996. HLA Management Plan: High Level Architecture for Modeling and Simulation. DMSO. Alexandria, VA.

Demaria, R. and Wilson, J. (2004). *High Score: The Illustrated history of electronic games*. New York: McGraw-Hill Publishing.

Dewar, R.D. & Dutton, J.E. (November, 1986). "The Adoption of radical and incremental innovations: An Empirical analysis". *Management Science*. 32(11), pp. 1422-1434.

Dodsworth, Clark. (1998). *Digital Illusion: Entertaining the Future with High Technology*. New York: ACM Press.

Dunnigan, J. (1993). *The Complete wargames handbook: How to play, design, and find them*. New York: William Morrow.

Elmaghraby, S.E. (June 1968). "The role of modeling in IE design". *Industrial Engineering*, 6, pp. 292-305.

Epstein, Joshua M. (1985). *The Calculus of Conventional War: Dynamic Analysis without Lanchester Theory*. Brookings Institution.

Federal Railroad Administration. (1997). "Basic Characteristics of Freight Rail Transportation in the United States". Report to the President's Commission on Critical Infrastructure Protection. Washington D.C.

Federation of American Scientists. (2006). *R&D challenges for games in learning*. Washington D.C.: Federation of American Scientists.

Federation of American Scientists. (2006). *Summit on educational games: Harnessing the power of video games for learning*. Washington D.C.: Federation of American Scientists.

Ferraris, Jonathan, (January, 2001). "Quadtrees", GameDev.net.

Ferren, Bran. (May-June 1999). "Some Brief Observations on the Future of Army Simulation". *Army RD&A Magazine*.

Feustel, Edward. 1995. Common Model of the Mission Space—Working Draft v.3. Modeling and Simulation Information System. Orlando, FL.

Fishwick, P. 1995. *Simulation Model Design and Execution*. Englewood Cliffs, NJ: Prentice Hall.

Foster, R. and Kaplan, S. (2001). *Creative destruction: Why companies that are built to last underperform the market—and how to successfully transform them*. New York: Currency Doubleday.

Frisken, S. & Perry, R. (May, 2003) "Simple and Efficient Traversal Methods for Quadtrees and Octrees", *Journal of Graphics Tools*, Vol. 7, Issue 3.

Fujimoto, R. 2000. *Parallel and Distributed Simulation Systems*. New York: Wiley-Interscience.

Fujimoto, Richard M. 1998. "Parallel and Distributed Simulation". *Handbook of Simulation: Principles, Methodology, Advances, Applications, and Practice*. John Wiley & Sons Inc. New York, NY.

Fujimoto, Richard. (1990). "Parallel Discrete Event Simulation". *Communications of the ACM*, Association of Computing Machinery.

Ghamari-Tabrizi, S. (1995). *The Worlds of Herman Kahn: The Intuitive science of thermonuclear war*. Boston, MA: Harvard University Press.

Gladwell, M. (2002). *The Tipping point: How little things can make a big difference*. New York: Little, Brown, & Co.

Hamm, S. (September 18, 2006). "Disaster management 101". *Business Week Magazine*, p. 12.

Hammer M. and Champy, J. (1993). *Reengineering the corporation: A Manifesto for business revolution*. Boston, MA: Harvard Business Review Press.

Hargadon, A. (2003). *How breakthroughs happen: The Surprising truth about how companies innovate*. Boston, MA: Harvard Business School Press.

Hart, S.L. and Milstein, M.B. (Fall 1999). "Global sustainability and the creative destruction of industries. *Sloan Management Review*, pp. 23-33.

Heinlein, Robert A. 1940. And He Built a Crooked House. 1983 reprint in *The Unpleasant Profession of Jonathan Hoag*. Berkley Books. New York, NY.

Herz, J. and Macedonia, M. (April 2002). "Computer games and the military: Two views". *Defense Horizons*, 11. Accessed at: http://www.ndu.edu/inss/DefHor/DH11/DH11.htm

Hof, R. (August 17, 2003). Andy Grove: "We can't even glimpse the potential".

Business Week Online. http://www.businessweek.com/@@GovuBoUQa-QmEPwkA/magazine/content/03_34/b3846612.htm

Hof, R. (May 1, 2006). "My Virtual Life". Business Week Online. Accessed at http://www.businessweek.com/magazine/content/06_18/b3982001.htm.

Hughes, Wayne P. (1997). *Military Modeling for Decision Making*. Military Operations Research Society. Alexandria, VA.

Huntington, S.P. (1996). *The clash of civilizations and the remaking of world order*. New York: Touchstone Press.

Hutchison, J. (Nov 1997). "The Junior Red Cross goes to Healthland". *American Journal of Public Health*, 87(11), pp.1816-1823.

IBM. (2006). "Expanding the innovation horizon: The Global CEO study 2006". IBM web site, accessed at http://www.ibm.com/innovation/ceo.Innovation

Janis, I.L. (1989) *Crucial decisions: Leadership in policymaking and crisis management*. New York: The Free Press.

Jennings, D. and Wattman, S. (1998). *Decision making: An Integrated approach*. London: Pitman Publishing.

Johnson, Jeff. (2004). "Massively-Multiplayer Engineering", *Proceedings of the Game Developers Conference*.

Jones, J.; Aguirre, D.; & Calderone, M. (September 2004). "The 10 Principles of change management". *Strategy+Business*, Issue 35.

Kelleghan, M. (July 1997). "Octree Partitioning Techniques", *Game Developer Magazine*.

Kim, I., et al. (2005). "Mall entertainment and shopping behaviors: A Graphical modeling approach". *Advances in Consumer Research*, 32.

Kim, W. and Mauborgne, R. (Jan-Feb 1999). "Creating new market space". *Harvard Business Review*.

Kim, W. and Mauborgne, R. (Sept-Oct 2000). "Knowing a winning business idea when you see one". *Harvard Business Review*.

Kimber, T. (2005). Professor's web site at State University of New York. http://people.morrisville.edu/~kimbert/beta_mws/pois17.gif

Knepell, P. and Arangno, D. 1993. *Simulation Validation: A Confidence Assessment Methodology*. Los Alamitos, CA: IEEE Press.

Konrad, R. (January 8, 2007). "IBM launches push into virtual world". *Business Week* Online. Accessed at http://www.businessweek.com/ap/financial-news/D8MHDIG80.htm?chan=search

Kuhl, F., Weatherley, R., and Dahmann, J. 2000. *Creating Computer Simulation Systems: An Introduction to the High Level Architecture*. Upper Saddle River, NJ: Prentice Hall PTR.

Kushner, D. (Aug 2002). "The wizardry of id". *IEEE Spectrum*, 39(8), pp.42-47.

Laird, J. (March 2000). Personal conversation with the author at the Game Developers Conference in San Jose, CA.

Lane, D. (May 1995). "On a Resurgence of management simulations and games". *Journal of the Operational Research Society*, 46(5), pp. 604-625.

Lane, J.L.; Slavin, S.; and Ziv, A. (June, 2001). Simulation in medical education: A Review. *Simulation & Gaming*, 32(297).

Law, A. and W. D. Kelton. 1991. *Simulation Modeling & Analysis*, 2nd edition. New York: McGraw-Hill.

Learning Federation, The. (2006). *Learning science and technology roadmap*. Washington D.C.: Federation of American Scientists.

LeHew, M. and Fairhurst, A. (June 2000). "US Shopping mall attributes: an exploratory investigation of their relationship to retail productivity". *International Journal of Retail and Distribution Management*, 28(6).

Leifer, R.; McDermott, C.; O'Connor, G.; Peters, L.; Rice, M.; & Veryzer, R. (2000). *Radical innovation: How mature companies can outsmart upstarts*. Boston: Harvard Business School Press.

Lenoir, T. (2003). "Programming theatres of war: Gamemakers as soldiers". In Latham, R. *Bombs and Bandwidth: The emerging relationship between information technology and security*. New York: The New Press. Draft chapter accessed at http://www.stanford.edu/dept/HPST/TimLenoir/Publications/Lenoir_TheatresOfWar.pdf

Leonard, D. (1995). *Wellsprings of Knowledge: Building and sustaining the sources of innovation*. Boston, MA: Harvard Business School Press.

Lewis, W.W., & Lawrence, H.L. (1990). "A new mission for corporate technology". *Sloan Management Review*, 31(4), pp. 57-67.

Lofdahl, Corey. (2002). "Characterizing the Terrorist Threat". Presentation at the 70th Military Operations Research Society Symposium. Leavenworth, Kansas.

Maier, F. and Grobler, A. (July 2000). "What are we talking about?—A Taxonomy of computer simulations to support learning". *System Dynamics Review*, 16(2), pp. 135-148.

Marsh, Robert. (October, 1997). "Critical Foundations: Protecting America's Infrastructure". Report of the President's Commission on Critical Infrastructure Protection. Washington D.C.

Mattern, Friedeman. (1993). "Efficient Algorithms for Distributed Snapshots and Global Virtual Time Approximation", *Journal of Parallel and Distributed Computing*. Vol 18, Num 4.

May, Janet O. (2002). "OneSAF Killer/Victim Scoreboard Capability for C2 Experimentation", *Proceedings of the 2002 Conference on Behavioral Representation in Modeling and Simulation*.

Mayo, M.; Singer, M.; & Kusumoto. (December 2005). "Massively multiplayer environments for asymmetric warfare". *Proceedings of the 2005*

McConnon, A. (October 16, 2006). "Virtual world, real courtroom". *Business Week Magazine*, p. 13.

McDaniels, B.A. (June 2005). "A Contemporary view of Joseph A. Schumpeter's theory of the entrepreneur". *Journal of Economic Issues*, 39(2), pp. 485-489.

McGarry, Stephen. 1995. The DoD High Level Architecture (HLA) Run-Time Infrastructure (RTI) and Its Relationship to Distributed Simulation. *Summary Report - 13th Workshop on Standards for the Interoperability of Distributed Simulations*. Institute for Simulation and Training. Orlando, FL.

McGrath, R.G. (2006). "Technology deployment and idiosyncratic competitive advantage". Knowledge @ Wharton online.

Medcof, J. and Yousofpourfard, H. (June 2006). "The Chief technology officer and organizational power and influence". International Association of Management of Technology Conference Proceedings.

Medeiros, Watson, Carson, & Manivannan. Editors. 1998. *1998 Winter Simulation Conference Proceedings*. Washington D.C.

Metcalfe, R. (June 2004). Why IT matters. *Technology Review*. Accessed at http://www.technologyreview.com/articles/04/06/metcalfe0604.asp?p=1

Michael, D and Chen, S. (2005). *Serious games: Games that educate, train, and inform*. Thompson Publishing.

Miller, D.C. and Thorpe, J.A. (August 1995). "SIMNET: The Advent of Simulator Networking". *Proceedings of the IEEE*. 83(8), pp. 1114-1123.

Moller, B and Lof, S. (Fall 2006). "A Management overview of the HLA evolved web service API". *Proceedings of the Fall 2006 Simulation Interoperability Workshop*.

Moore, G.A. (1991). *Crossing the chasm: Marketing and selling technology products to mainstream customers*. New York: Harper Business.

Moore, G.A. (1995). *Inside the tornado: Marketing strategies from Silicon Valley's cutting edge*. New York: Harper Business.

Moore, G.A. (2005). *Dealing with Darwin: How great companies innovate at every phase of their evolution*. New York: Portfolio Books.

Morris, C. and Ferguson, C. (March-April 1993). "How architecture wins technology wars". *Harvard Business Review*. Boston, MA: Harvard Business School Press.

Morse, P.M. and Kimball, G.E. 1998. *Methods of Operations Research*. Alexandria, Virginia: Military Operations Research Society.

Mowbray, Thomas J. and Zahavi, Ron. 1995. *The Essential CORBA: System Integration Using Distributed Objects*. John Wiley & Sons Inc. New York, NY.

Musgrove, M. (August 30, 2006). "A Computer game for real-life crises". *Washington Post*.

Nalebuff, B. & Brandenburger, A. (Nov/Dec 1997). "Co-opetition: Competitive and cooperative business strategies for the digital economy". *Strategy & Leadership*, 25(6).

Nance, R. 1996. "A History of Discrete Event Simulation Programming Languages". *The History of Programming Languages - II*. New York: Association of Computing Machinery.

National Research Council. (1997). *Modeling and simulation: Linking entertainment and defense*. Washington DC: National Academies Press.

National Research Council. (2006). *Defense modeling, simulation, and analysis*. Washington D.C.: National Academies Press.

National Simulation Center. 2000. *Training with Simulations: A Handbook for Commanders and Trainers*. Fort Leavenworth, KS.

Naval Training Systems Center. (1990). *Foundations for the future*. Navy Publication.

Nelson, Jennifer. (2002). "Critical Infrastructure Surety Activities at Sandia National Laboratories". http://www.sandia.gov/

Nie, N. and Erbring, L. (Summer 2002). "Internet and society: A Preliminary report". *IT & Society*, 1(1), pp.275-283.

Nolan, R. L. (1992). "The Stages Theory: A Framework for IT Adoption and Organizational Learning." In *America's Information Technology Agenda*, edited by J. Mechling and C. Rosenberg. Cambridge: John F. Kennedy School of Government.

O'Neill, P.H., & Bridenbaugh, P.R. (November-December, 1992). Credibility between CEO and CTO—A CEO's perspective; Credibility between CEO and CTO – A CTO perspective. *Research Technology Management*, 35(6), 25-34.

Orbanes, P.E. (2004). *The Game makers: The Story of Parker Brothers*. Boston: Harvard Business School Press.

Parry, Samuel, Editor. (1995). *Military OR Analyst's Handbook: Conventional Weapons Effects*. Military Operations Research Society.

Perelman, M. (Summer 1995). "Retrospectives: Schumpeter, David Wells, and Creative Destruction". *Journal of Economic Perspectives*, 9(3), pp.189-197.

Perla, P. 1990. *The Art of Wargaming: A Guide for Professionals and Hobbyists*. Annapolis, MD: Naval Institute Press.

Perumalla, Kalyan. (July 2004). "libSynk: Source Code for Time Synchronization", available online at
http://www.cc.gatech.edu/computing/pads/kalyan/libsynk.htm.

Plato. 1937. "The Republic". *The Dialogues of Plato*. Translated by B. Jowett. Random House. New York, NY. (Originally written circa 380 B.C.)

Porter, M. (March 2001). "Strategy and the Internet". *Harvard Business Review*.

Prahalad, C.K. (2006). "The Innovation sandbox". *Strategy+Business*, Issue 44.

Prensky, M. (2001). *Digital game-based learning*. New York: McGraw Hill.

Prensky, M. (2006). *Don't bother me mom—I'm learning: How computer and video games are preparing your kids for 21st century success—and how you can help*. St. Paul, MN: Paragon House.

Pritchard, M. (2001). "A High-Performance Tile-based Line-of-Sight and Search System", *Game Programming Gems 2*.

Pritsker, A. 1990. *Papers, Experiences, Perspectives*. West Lafayette, IN: Systems Publishing Corp.

Rashid, Ahmed. (2001). *Taliban: Militant Islam, Oil & Fundamentalism in Central Asia*. Yale University Press. New Haven: CT.

Reinganum, J.F. (February 1985). "Innovation and industry evolution". *The Quarterly Journal of Economics*, pp. 81-99.

Rich, B. and Janos, L. (1996). *Skunk Works*. Little, Brown & Company.

Robinson, C., Woodard, J., & Varando, S. (Fall 1998). "Critical Infrastructure: Interlinked and Vulnerable". Issues in Science and Technology Online. http://www.nap.edu/issues/15.1/robins.htm

Rosen, L. and Weil, M. (June 1995). "Adult and teenage use of consumer, business, and entertainment technology: Potholes on the information superhighway". *Journal of Consumer Affairs*, 29(5), pp.55.

Rosner, R. (September 7, 2006). Keynote Address to the High Performance Computer Users Conference, Washington D.C.
http://www.hpcusersconference.com/agenda.html

Rowland, K. (August 30, 2006). "Computer game will train first responders". *Washington Times*.

Rucker, Rudy. 1984. *The Fourth Dimension: A Guided Tour of the Higher Universes*. Houghton Mifflin Company. Boston, MA.

Russell, S. and P. Norvig. 1995. *Artificial Intelligence: A Modern Approach*. Reading, MA: Prentice Hall.

Sandia National Laboratories. (1997). "US Infrastructure Assurance Prosperity Game Final Report". Report to the President's Commission on Critical Infrastructure Protection. Washington D.C.

Sanson, J.P. 2006. Studying War: Central Louisiana and the Training of United States Troops 1939-1945. Accessed at http://www.crt.state.la.us/tourism/lawwii/maneuvers/Studying_War.htm

Scacchi, W. (Jan-Feb, 2004). "Free and open source development practices in the game community". *IEEE Software*, 21(1), pp. 59-66.

Schilling, M. (Spring 2003). "Technological leapfrogging: Lessons from the U.S. video game console industry". *California Management Review*, 45(3).

Schrage, M. (2000). *Serious play: How the world's best companies simulate to innovate*. Boston: Harvard Business School Press.

Schriber, T. 1991. *An Introduction to Simulation Using GPSS/H*. New York: John Wiley.

Schumpeter, J.A. (1942). *Capitalism, Socialism, and Democracy*. New York: Harper & Row Publishers.

Shankar, V. and Bayus, B. (2003). "Network effects and competition: An Empirical analysis of the home video game industry". *Strategic Management Journal*.

Shankland, S. (December 7, 2006). "The world needs only five computers —Interview with Greg Popadopoulos". C|Net News. Accessed at http://news.com.com/The+world+needs+only+five+computers/2008-1011_3-6141598.html

Sheff, D. (1999). *Game over: Press start to continue*. Wilton, CT: Cyber Active Publishing.

Shenk, D. (2007). *The Immortal game: A History of Chess*. New York: Anchor Books.

Shubik, Martin, Editor. (1983). *Mathematics of Conflict*. Elsevier Science Publishers.

Shuster, L. (June 2003). "Global gaming industry now a whopping $35 billion market". Compiler Online.
http://www.synopsys.com/news/pubs/compiler/art1lead_nokia-jul03.html

Sieberg, D. (November 23, 2001). "War games: Military training goes high-tech". CNN Sci-Tech Online, accessed at http://archives.cnn.com/2001/TECH/ptech/11/22/war.games/index.html

Singhal, S. and M. Zyda. 1999. *Networked Virtual Environments: Design and Implementation*. New York: ACM Press.

Smed, J.; Kaukoranta, T. & Hakonen, H. (April 2002). "A Review on networking and multiplayer computer games". Technical Paper TR454 from the Turku Centre for Computer Science, University of Turku, Finland. Accessed at http://staff.cs.utu.fi/~jounsmed/papers/TR454.pdf

Smith, D. and Alexander, R. (1999). *Fumbling the future: How Xerox invented then ignored the first personal computer*. New York: HarperCollins.

Smith, K.G.; Ferner, W.J.; & Grimm, C.M. (May 2001). "King of the hill: Dethroning the industry leader". *Academy of Management Executive*, 15(2), pp. 59-70.

Smith, R. (December 2008). "Game Impact Theory: The Five Forces That are Driving the Adoption of Game Technologies within Multiple Established Industries". *Proceedings of the 2008 Interservice/Industry Training, Simulation, and Education Conference*.

Smith, R. (Jan-Feb 2007). "The Disruptive Potential of Game Technologies: Lessons Learned from its Impact on the Military Simulation Industry". *Research Technology Management*, 50(1).

Smith, R. (January 2006). "Technology disruption in the simulation industry". *Journal of Defense Modeling and Simulation*, 3(1), pp. 3-10.

Smith, R. (July-August, 2003). "The Chief Technology Officer: Strategic Responsibilities and Relationships". *Research-Technology Management*.

Smith, R. (Sept-Oct 2006). "The Disruptive Potential of Game Technologies: Lessons Learned from its Impact on the Military Simulation Industry". *Research Technology Management*, 49(5).

Smith, R. 2000. Simulation. In *Encyclopedia of Computer Science*, 4th edition, edited by Ralston, Reilly, and Hemmendinger, (1578-1587). New York: Grove's Dictionaries.

Smith, Roger D. 1995. Techniques in Simulation Interoperability. *1995 Georgia Tech Course on Modeling, Simulation, and Gaming of Warfare*. Georgia Tech Research Institute. Atlanta, GA.

Smith, Roger D. 1995. The Conflict Between Heterogeneous Simulations and Interoperability. *Proceedings of the 1995 Interservice / Industry Training Systems and Education Conference*. Albuquerque, NM.

Smith, Roger D. 1996. *Proceedings of the Electronic Conference on Interoperability in Training Simulation*. Elecsim96.

Solow, R. (August 1957). "Technical change and the aggregate production function". *Review of Economics and Statistics*, 39(3), pp. 312-320.

Spiers, D. (July 5, 2001). "CTOs: Technology's easy—It's the people part that's hard to master". *Business 2.0*. http://www.business2.com/b2/subscribers/articles/0,17863,513730,00.html

Squire, K. and Steinkuehler, C. (2007). "Generating cyberculture/s: The Case of Star Wars Galaxies". In D. Gibbs and K.L. Krause (Eds.), *Cyberlines 2.0 languages and cultures of the Internet*. Albert Park, Australia: James Nicholas Publishers.

Steinkuehler, C. (2005). "The new third place: Massively multiplayer online gaming in american youth culture". *Tidskrift Journal of Research in Teacher Education*, 3, pp.17-32. Umea Universitet (Sweden).

Steinkuehler, C. (January 2007). "Massively multilayer online video gaming as participation in a discourse". *Mind, Culture, and Activity*, 13(1), pp.38-52.

Stephenson, N. (1992). *Snow Crash*. Bantam Books.

Sterman, John. (2000). *Business Dynamics: Systems Thinking and Modeling for a Complex World*. McGraw Hill. New York: NY.

Stokes, D.E. (1997). *Pasteur's Quadrant: Basic science and technology innovation*. Washington DC: Brookings Institution Press.

Strnadl, C. (Fall 2006). "Aligning business and IT: The Process-driven architecture model". *Information Systems Management*, pp. 67-77.

Svarovsky, J. (2000). "Multi-Resolution Maps for Interaction Detection", *Game Programming Gems*, Charles River Media.

Swindler, J. (November 2006). "Coach for a day". *Fast Company Magazine*, p.34.

Thomke, S. (2003). *Experimentation matters: Unlocking the potential of new technologies for innovation*. Boston, MA: Harvard Business School Press.

Tracz, Will. 1995. *Confessions of a Used Program Salesman: Institutionalizing Software Reuse*. Addison-Wesley Publishing company. Reading, MA.

Tushman, M. & O'Reilly, C. (1997). *Winning through innovation: A Practical guide to leading organizational change and renewal*. Boston, MA: Harvard Business School Press.

U.S. Army. (1990). *Field Artillery Handbook*. U.S. Army, 1990.

U.S. Government. (2001). "National Plan for Information System Protection, Version 1.0". Washington D.C.

Uleman, R. (June 2006). "Service oriented architecture unveiled". *Geospatial Solution*.

Unger, Brian & Ferscha, Alois. Editors. (1998). *Proceedings of the Twelfth Workshop on Parallel and Distributed Simulation*. Banff, Alberta.

Utterback, J. (1996). *Mastering the dynamics of innovation*. Boston, MA: Harvard Business School Press.

Venkatraman, N. & Lee, C. (Dec 2004). "Preferential linkage and network evolution: A Conceptual model and empirical test in the U.S. video game sector". *Academy of Management Journal*, 47(6).

von Hippel, E. (2002). "Innovation by user communities: Learning for open-source software", in *Innovation: Driving product, process, and market change.* ed. E. Roberts. San Francisco, CA: Jossey-Bass.

von Hippel, E. (2005). *Democratizing innovation.* Boston, MA: MIT Press.

von Hippel, E. (Summer 2001). "Innovation by user communities: Learning from open-source software". *MIT Sloan Management Review.*

Weiner, M. (1959). *An introduction to war games.* RAND publication P-1773. Available online at http://www.rand.org/pubs/papers/P1773/.

Wharton Business School. (February 27, 2006). "Ben Franklin forum on innovation: What can you learn from the world's top innovators?" Knowledge @ Wharton online.

Wikipedia. (2007). "List of United States Army MOS". http://en.wikipedia.org/wiki/List_of_United_States_Army_MOS

Williams, O. (January 2007). Conversations between the CTO and CIO offices of U.S. Army PEO-STRI.

Wilson, T. (October 2006). "Top 20 game publishers". *Game Developer Magazine*, 13(9), pp.12-22.

Winkler, T and Buckner, K. (Fall 2006). „Receptiveness of gamers to embedded brand messages in advergames: Attitudes toward product placement". *Journal of Interactive Advertising*, 7(1), pp.37-46.

Yu, Alan. Editor. (1999). *Proceedings of the 1999 Game Developers Conference.* San Jose, California.

Yunus, M.; Mallal, R.; and Shaffer, D. (Jan 14, 2007). "Amazon EC2 and Oracle SOA Suite a strong combo". *Dr. Dobb's Journal.*

Zachman, J. (1987). "A Framework for information systems architecture". *IBM Systems Journal*, 26(3), pp. 454-470.

Zeigler, Bernard P. (1976). *Theory of Modelling and Simulation.* Robert E. Krieger Publishing Company. Malabar, FL.

Zyda, M. (June 2006). "Educating the next generation of game developers". *IEEE Computer.*

Zyda, M. (September 2005). "From visual simulation to virtual reality to games". *IEEE Computer*, 38(9), 30-34.

Zyda, M., et.al. (December 2003). "This year in the MOVES institute". *Proceedings of the Second International Conference on Cyberworlds*. pp. xxxii.